An Introduction to
Particle Dark Matter

Advanced Textbooks in Physics

ISSN: 2059-7711

Published

Trapped Charged Particles: A Graduate Textbook with Problems and Solutions
edited by Richard C Thompson, Niels Madsen & Martina Knoop

Studying Distant Galaxies: A Handbook of Methods and Analyses
by François Hammer, Mathieu Puech, Hector Flores & Myriam Rodrigues

An Introduction to Particle Dark Matter
by Stefano Profumo

Forthcoming

An Introducion to String Theory and D-Brane Dynamics: With Problems and Solutions (3rd Edition)
by Richard J Szabo

Advanced Textbooks in Physics

An Introduction to Particle Dark Matter

Stefano Profumo
UC Santa Cruz & Santa Cruz Institute for Particle Physics, USA

NEW JERSEY • LONDON • SINGAPORE • BEIJING • SHANGHAI • HONG KONG • TAIPEI • CHENNAI • TOKYO

Published by

World Scientific Publishing Europe Ltd.

57 Shelton Street, Covent Garden, London WC2H 9HE

Head office: 5 Toh Tuck Link, Singapore 596224

USA office: 27 Warren Street, Suite 401-402, Hackensack, NJ 07601

Library of Congress Cataloging-in-Publication Data
Names: Profumo, Stefano, 1978– author.
Title: An introduction to particle dark matter / by Stefano Profumo (UC Santa Cruz & Santa Cruz Institute for Particle Physics, USA).
Description: Hackensack, NJ : World Scientific, [2017] | Series: Advanced textbooks in physics
Identifiers: LCCN 2016041680| ISBN 9781786340009 (hc ; alk. paper) | ISBN 1786340003 (hc ; alk. paper) | ISBN 9781786340016 (pbk ; alk. paper) | ISBN 1786340011 (pbk ; alk. paper)
Subjects: LCSH: Dark matter (Astronomy) | Particles (Nuclear physics)
Classification: LCC QB791.3 .P756 2017 | DDC 523.1/126--dc23
LC record available at https://lccn.loc.gov/2016041680

British Library Cataloguing-in-Publication Data
A catalogue record for this book is available from the British Library.

Desk Editors: Anthony Alexander/Mary Simpson

Typeset by Stallion Press
Email: enquiries@stallionpress.com

Printed in Singapore

Preface

The discovery of dark matter as a particle might be just around the corner. I am convinced that such discovery will come from a combination of:

(1) the *ingenuity* and "out-of-the-box" thinking of theorists and experimentalists alike, and of
(2) the ability to master the *wisdom* we accumulated about how to model and how to search for dark matter as a particle.

This book is not about item (1). There are many ways to train and foster creative thinking, such as cycling, yoga, surfing, and other more or less legal activities. And, of course, the practice of physics. This book is, instead, a modest attempt at helping with item (2).

Like for any other major open intellectual problem, people have developed a variety of techniques to address the question of the nature of dark matter. Learning about this "bag of tricks" allows one to think on a higher level than their predecessors, and thus to *"stand on the shoulders of giants"*, as Newton famously said. This book collects lessons I personally learned and used in my research work that I think are valuable tools in that bag.

What follows is not a review. At times I found myself unable to derive with self-consistency and brevity a result worth being aware of. In those cases, indeed, I simply quoted and reviewed that result, pointing the Reader to the relevant literature — as one would find in any review. However, as a rule, in this book I try to explain matters, and to derive them, as if I were to introduce them to a smart but not necessarily knowledgeable audience at the blackboard. This *"blackboard style"* is something I learned from my mentor Marc Kamionkwski at his group meetings while we were both at Caltech, as well as from the many talented colleagues who manage to quickly and clearly explain the exciting pieces of physics that keep them busy and excited, without assuming extensive prior knowledge.

The choice of topics in this book has chiefly to do with my personal taste and bias. Again, this is not meant to be a comprehensive review on particle dark matter. Rather, I cherry-pick results and "stories" that I believe have some intrinsic value for the Reader. Sometimes these lessons are simply beautiful pieces of physics; often they are tools of the "bag of tricks" that I consider too important to pass. The book also contains two appendices with a review of topics and results in both introductory cosmology and basic particle physics, necessary background for most of the discussion.

I apologize for often not quoting the *historically original* references: this is not a *history* of particle dark matter. References reflect my own personal intellectual trajectory: I cite the sources I myself utilize to learn about a given topic.

My personal taste and *modus operandi* also transpires from the choice to, occasionally, go in depth on certain topics that have captured my curiosity. At times, I realize, these excursions might not appear warranted or necessary, but they reflect how I like doing research: I get passionate about a subject, and I dig deep into it. It goes without saying that the Reader should feel free to skip such diversions should she not share my curiosity on the topic![a]

A key tool I use in what follows is plenty of *exercises* (more than 200) that I strongly encourage the Reader to work through. In some cases the exercises are about the detailed derivation of results I quote in the discussion, in other cases they are good intellectual gymnastics. My hope is that every exercise serves the purpose of adding to the "bag of tricks", or to the ability of putting those tricks to good use. One of my best-ever undergraduate students once complained that I did not give enough homework: that was the main way for him to learn the (graduate-level) material I discussed in class. Hopefully he is satisfied with the quality, quantity, and purpose of the exercises proposed here.

At a recent lecture series on dark matter, at my asking whether my lectures were clear, a student told me that she was "sometimes lost" and that it was "entirely her fault". I immediately told her that it is never the student's fault if a lecturer fails at getting the science through — it is always the instructor's fault! I have benefited immensely from the sometimes harsh

[a]This is mentioned as right #2 in Daniel Pennac's "The Rights of the Reader". Right #3 is the right "not to finish a book", #4 to "read it again": they all apply here...

but well-meaning feedback I received from my students on this book, and I am grateful and very much indebted to them. The best compliment I received on the material presented here, in some early incarnation, was from a student who after one of my research group meetings wrote me that he *"was pleasantly surprised that reading the book was actually really useful to understanding what was being discussed at the meetings"*. It does not get much better than being told that what you teach is useful.

The kernel of the material presented in this book originates from a series of four lectures on astrophysical probes of dark matter I gave in June 2012 at the Theoretical Advanced Study Institute in Elementary Particle Physics (TASI) Summer School at the University of Colorado, Boulder. The aim of that set of lectures was neither to present in detail particle dark matter models, nor to focus on the technical aspects of experiments or the (alas, necessary!) understanding of astrophysical backgrounds; what I intended to do back then was to try to convey and to work out order-of-magnitude estimates and physics lessons that could be applied to a variety of particle dark matter models and physical situations.

Since then I have had the pleasure and opportunity to expand the purpose of that original set of lectures, and to present the new material at lectures on particle dark matter on various occasions (at TAUP 2013, at the 2013 pre-SUSY summer school, at the Galileo Galilei Institute in 2016, at the 2016 pre-SUSY school, and at schools in Brazil, Korea, and Mexico). The level and natural audience for this book is the same as it was for my lectures: advanced undergraduate and graduate physics or astronomy students. However, I believe that others might also find my discussion of some use: scholars with expertise in other fields interested in learning about dark matter as a particle; astronomers curious about the particle physics of dark matter; or lay persons keen on appreciating how theorists think about the question of the fundamental nature of dark matter. I welcome input and feedback from all Readers, and on any aspect of the book — choice of topics, style, exercises, references, etc. I hope you enjoy what follows.

About the Author

Stefano Profumo is a Professor of Physics at the University of California, Santa Cruz, the Deputy Director for Theory at the Santa Cruz Institute for Particle Physics, and the Director of Graduate Studies for the Physics PhD Program.

Profumo earned his MS from Scuola Normale Superiore in Pisa, Italy, in 2001, and a PhD in Elementary Particles Theory from the International School for Advanced Studies in Trieste, Italy, in 2004. He joined the faculty at the University of California in 2007 after two postdoctoral positions at Florida State University (2004–2005) and at the California Institute of Technology (2005–2007).

Profumo's research work spans particle physics theory, cosmology, and high-energy astrophysics. He has published over 140 peer-reviewed articles, many of which are on the topic of particle dark matter; his research work has been cited more than 10,000 times, according to Google Scholar.

Acknowledgments

I would like to thank Elena Pierpaoli for inviting me to lecture at TASI 2012. It was fun, and it was great to meet so many, promising young fellas at that school and at the many other schools I had the privilege to lecture at. Thanks to Eric Carlson, Jonathan Cornell, Kfir Dolev (the smart undergraduate), Nicolas Fernandez, Jonathan Kozaczuk, Tim Linden, and Jakub Scholtz for carefully proofreading early versions of this book and for their feedback. Special thanks to Francesco D'Eramo, who scrupulously read through the entire manuscript, carried out all of the exercises, and made sure that everything contained in this book is correct (if not, it is his fault, not mine). Francesco's feedback was truly invaluable, and *scherzi a parte*, I really am very grateful to him, a true expert and scholar on most matters discussed in this book.

My research is partly supported, and has been supported in part, by the Department of Energy, the National Science Foundation, and NASA.

Contents

Chapter 1

Particle Dark Matter: The Name of the Game

1.1 Praeludium

In a letter to Sommerfeld dated December 9, 1915, Albert Einstein stated: "How helpful to us is Astronomy's *pedantic accuracy*, which I used to ridicule!" (my emphasis).[a] He was referring to measurements of the advance of the perihelion of Mercury, one of the key observations testing predictions of General Relativity. The discovery of dark matter is a history of *pedantic astronomical observations*, leading to one coherent picture of a preposterous universe. One where only one fifth of the matter content is made of particles we know of, the rest being something we fundamentally know very little about: dark matter. And I am convinced that it will be thanks to the same, persistent *pedantic accuracy* of astronomers and particle physics experimentalists that we will eventually write the chapter of the book of physics about the nature of dark matter.

Thus far, we have been able to learn what the dark matter, as a particle, is not, by cleverly looking for its possible *microscopic* manifestations. However, we do know a fair amount about dark matter at a *macroscopic* level, and how such macroscopic features are connected with particle properties of the dark matter. We know very accurately how much dark matter is out there, globally, in the universe, and how the dark matter is distributed in selected regions of the universe; we quantitatively know that, besides gravity, dark matter interacts weakly, if at all, with the particles (of the Standard Model (SM) of particle physics) we know and love; we know that dark matter is cold, or at best warm (in a sense to be made clear later on);

[a]Quote credit to a slide in a seminar by Rocky Kolb.

finally, we know that dark matter has been out there for a long while and still is — this implies that, as a particle, the dark matter must be stable, or very long lived.

Learning how we convinced ourselves that dark matter is indeed filling up the universe is a great way to learn facts that can be used to build particle models for the dark matter. This is the *raison d'être* for this chapter, besides perhaps giving you some ammunitions to explain what you do for a living to the inevitable chatty guy sitting next to you on a plane.

1.2 There is more matter than the matter that shines — classical (mechanical) evidences

Many good reviews exist on evidences for dark matter (for a historical perspective see e.g. the recent book [1]). As I explained in the Preface, this book is not a review. Rather, here I choose to present a few select stories that feature interesting pieces of physics and that contain "life lessons" about dark matter as a particle.

Zwicky and the virial theorem

When you attend a seminar on dark matter, chances are the speaker will produce a one-liner about Zwicky's 1930s "discovery" of dark matter in the Coma cluster, possibly accompanied by a funny picture of Zwicky doing the OK sign and a weird face (Fig. 1.1). If the Reader contents herself with that one-liner and funny picture, she should feel free to skip to the next section.

Figure 1.1: The funny guy who, some say, invented the name "dark matter".

If not, here is a somewhat quantitive account of what Zwicky actually did and said.

I think it is important to dig into Zwicky's original arguments first because of history (a history that not many know well); second because classical mechanics is really beautiful, and Zwicky's paper features a few little gems; and third because, as anything good in physics, Zwicky's are simple arguments that only require you to know one equation: $F = ma$.

The logic that led Fritz Zwicky to his visionary statement that "should this turn out to be true, the surprising result would follow that dark matter is present in a much higher density than radiating matter" [2] is simple: the virial theorem applied to the motion of galaxies (or "nebulae" as they were called back then — it had not been long since people had figured out that they were extragalactic objects and not clouds (*nebulae*, in Latin) of gas or dust in the galaxy) in a galaxy cluster (specifically, Coma).

Let's consider a "nebula" i at position $\vec{r}_i$ and of mass M_i, and take the scalar product of "$F = ma$" with $\vec{r}_i$:

$$\vec{r}_i \cdot \left(M_i \frac{d^2\vec{r}_i}{dt^2} = \vec{F}_i \right). \tag{1.1}$$

Now let us sum Eq. (1.1) over i, i.e. over all "nebulae" in the cluster, and get

$$\frac{1}{2}\frac{d^2\Theta}{dt^2} = \mathrm{Vir} + 2K_T,$$

where

$$\Theta \equiv \sum_i M_i r_i^2$$

is the *polar moment of inertia*,

$$\mathrm{Vir} \equiv \sum_i \vec{r}_i \cdot \vec{F}_i$$

is the "*virial*" of the cluster, and K_T is the total nebulae kir
the cluster is *stationary*, the polar moment of inertia fluctu
constant value, so the time average (which we will indicate

its *time derivative* vanishes.[b] As a result, we get the *virial theorem*:

$$\overline{\text{Vir}} = -2\overline{K_T}.$$

Zwicky continues, rather prophetically, stating that "On the *assumption*[c] that Newton's inverse square law accurately describes the gravitational interactions among nebulae[d]", one gets

$$\text{Vir} = U = -\sum_{i<j} G_N \frac{M_i M_j}{r_{ij}}, \tag{1.2}$$

with G_N Newton's gravitational constant, and $r_{ij} \equiv |\vec{r}_j - \vec{r}_i|$.

Exercise 1. Prove Eq. (1.2); it might be helpful to use the fact that

$$\sum_{i=1}^{N} \vec{F}_i \cdot \vec{r}_i = \sum_{i=2}^{N} \sum_{j=1}^{i-1} \vec{F}_{ij} \cdot (\vec{r}_j - \vec{r}_i)$$

for a force that is a function only of the relative distance between particles.

The virial theorem then looks like

$$-\overline{U} = 2\overline{K_T} = \overline{\sum_i M_i v_i^2} = \sum_i M_i \overline{v_i^2}.$$

If one assumes uniform distribution of the cluster total mass M_{tot} across a sphere of radius R_{tot}, one gets

$$U = -G_N \frac{3M_{\text{tot}}^2}{5R_{\text{tot}}}. \tag{1.3}$$

[b]Another good reason for the average polar moment of inertia to vanish at large times is that it is bounded from above by the ratio of its maximal value and the time over which the average is taken; the maximal value is finite for a bound system, the ratio goes to zero for long-enough time averages.
[c]My emphasis.
[d]We will discuss modified theories of gravity in Sec. 1.6.

> **Exercise 2.** Prove Eq. (1.3), i.e. that the potential energy for a uniform sphere of radius $R_{\rm tot}$ and mass $M_{\rm tot}$ is what appears in the equation.

Taking an average not only over time, but also over nebulae velocity (second bar, following Zwicky's original notation),

$$\sum_i M_i \overline{v_i^2} = M_{\rm tot} \overline{\overline{v^2}},$$

which gives a proxy for the total mass

$$M_{\rm tot} = \frac{5 R_{\rm tot} \overline{\overline{v^2}}}{3 G_N}. \tag{1.4}$$

> **Exercise 3.** In the original paper, Zwicky points out that Eq. (1.4) can be derived in an alternate way: take the time average of Eq. (1.1) for an individual nebula, and again assume a stationary system. You will then get a version of the virial theorem for an individual nebula. Now consider the forces acting on said nebula from a sphere of mass $M_{\rm tot}$ and radius $R_{\rm tot}$, calculate the virial for the individual nebula $\overline{{\rm Vir}}_i$, and find an expression for $\overline{v_i^2}$ in terms of $\overline{r_i^2}$. Using the fact that we are assuming a uniform distribution, and therefore that the average nebula spends equal times in equal volumes, calculate the time-averaged $\overline{\overline{r_i^2}}$ and show that Eq. (1.4) follows.

> **Exercise 4.** Calculate how Eq. (1.4) changes if all mass were concentrated in two or three nebulae of mass $M_{\rm tot}/2$ or $M_{\rm tot}/3$, and these masses were at mutual distances as small as $R_{\rm tot}/10$. Show that
>
> $$U = -\frac{5}{2} G_N \frac{M_{\rm tot}^2}{R_{\rm tot}} \quad \text{and} \quad U = -\frac{10}{3} G_N \frac{M_{\rm tot}^2}{R_{\rm tot}},$$
>
> respectively.

After carrying out the exercise above, Zwicky concluded that, conservatively (meaning, according to Zwicky, and somewhat arbitrarily, using $U = -5G_N \frac{M_{\rm tot}^2}{R_{\rm tot}}$) one would get

$$M_{\rm tot} \gtrsim \frac{R_{\rm tot}\overline{\overline{v^2}}}{5G_N},$$

and proceeded substituting for the observed average velocity squared along the line of sight,

$$\overline{\overline{v_s^2}} \approx 5 \times 10^{15}\ {\rm cm}^2{\rm s}^{-2}.$$

Of course, $\overline{\overline{v_s^2}} = \overline{\overline{v^2}}/3$, therefore $M_{\rm tot} > 9 \times 10^{43}$ kg, or about $4.5 \times 10^{13}\ M_\odot$ (this latter symbol indicating one solar mass).

Exercise 5. Let us start warming up to big numbers and units conversion: check Zwicky's calculation of $M_{\rm tot}$.

Given that a typical nebula was thought at the time, from observations, to contain about 8.5×10^7 sun-like stars, Zwicky concluded with the "somewhat unexpected" result that the mass-to-light ratio in Coma is around 500 and (in the original German version of the article, not in the *Astrophysical Journal*) made famous the term "dark matter" ("*dunkel materie*") to indicate the missing mass in the system.[e] Interestingly, despite a variety of issues in Zwicky's estimates (including the distance to Coma, and therefore its physical size), we now know that the mass-to-light ratio of clusters of galaxies indeed asymptotes (for larger and larger systems) at around 400. Incidentally, this is a way to measure the total matter content of the universe, which gives $\Omega_m = \rho_m/\rho_{\rm crit} \simeq 0.3$, in accordance with other methods (see Appendix B for the relevant definitions, and see the discussion below for a list of alternate ways to determine Ω_m).

[e]Of course Zwicky did not "invent" dark matter; who did that is of little importance here.

Exercise 6. In his calculation, Zwicky used a value for the Hubble parameter (for a definition of the Hubble parameter see Appendix B) of 558 km/s/Mpc, while the central value today, from cosmic microwave background (CMB) data, is $\sim$67 km/s/Mpc [3]. Estimate the impact on the total cluster mass estimate of the "wrong" Hubble parameter. What went wrong in the opposite direction for Zwicky's bottom line to end up in the right ballpark? (Note that there is a significant discrepancy between the CMB-inferred value and the Hubble Space Telescope inferred value of 73.2 km/s/Mpc [4] — this discrepancy could be an interesting hint for a "new" cosmological species providing additional "dark radiation" but the jury is still out).

Exercise 7. If indeed globally the mass-to-light ratio is around 400, estimate how many sun-like stars are there in the universe.

Why, as 21st century particle physicists or cosmologists, do we care about Zwicky's old paper? Primarily because methods similar to those Zwicky employed (but of course with much better cosmological inputs and observations!) are still used to study the dark matter content, its velocity and density profile, in a wide range of astrophysical systems, ranging from dwarf galaxies up to clusters. Understanding the details of such systems might cue us on *particle* model building: for example, the absence or not of cores in dwarf spheroidal galaxies might have to do with a mass scale associated with the dark matter particle; the size of the dark matter halos impacts whether or not there is a small-scale structure problem that also might have to do with the kinetic decoupling of dark matter in the early universe (and thus with its microscopic properties), the detailed dynamics of halos inside halos (e.g. galaxies within a cluster) might constrain self-interactions of dark matter as a particle, etc. *We stand on the shoulder of giants!*

Rubin and galactic rotation curves

With the advent of radio telescopes, it became possible to measure the velocity, as a function of radius, of gas circling around cylindrically symmetric systems such as spiral galaxies. Typically, observations were made using the Doppler shift of the 21 cm hyperfine transition of neutral hydrogen. The expectation for the circular velocity[f] v_c for an axisymmetric galaxy as a function of the distance r from the axis of the galaxy (indirectly measured by the line redshift or blueshift, with some assumption about the gas density distribution as a function of radius) is simple:

$$v_c^2(r) = \frac{G_N M(r)}{r}, \tag{1.5}$$

where $M(r)$ is the total mass enclosed in a sphere of radius r. The circular speed, therefore, is a direct proxy for the total mass inside a radius r. The gravitational potentials for a few simple systems give instructive expected asymptotic behaviors for $v(r)$. Obviously, for a point mass M (and, by Newton's second theorem on the gravitational potential of spherical systems — see Exercise 8 — outside any spherically symmetric mass distribution of total mass M)

$$v_c(r) = \sqrt{\frac{G_N M}{r}}.$$

Exercise 8. Prove Newton's second theorem: "*The gravitational force on a body outside a closed spherical shell of matter is the same as it would be if all the shell's matter were concentrated into a point at its center*". Note that the proof of this theorem eluded Newton himself for several years, but it actually is not that hard. In case you feel an inferiority complex for Newton, open Binney and Tremaine's classic book [5] and enjoy.

The $v_c \sim r^{-1/2}$ behavior is known as "Keplerian" fall-off (Kepler was the first to determine that that behavior applies to the planetary motions in the solar system).

The simplest sketch for a galaxy is a homogeneous sphere of radius R and constant density ρ (and thus $M(r) = \frac{4}{3}\pi\rho r^3$, for $r < R$). In this case the circular velocity is *linear* within the galaxy,

$$v_c(r) = \sqrt{\frac{4\pi G_N \rho}{3}}\, r, \quad (r \leq R)$$

and, by virtue of what we discussed above, $v_c(r) \sim r^{-1/2}$ for $r > R$.

---

[f]That is, the speed of a test particle in a circular orbit.

Exercise 9. Show that for a homogeneous sphere of density ρ the circular orbital *period* T is independent of radius, and that the time it takes for a test particle released from rest at a radius r to reach $r = 0$ is a time $T/4$.

Exercise 9, incidentally, illustrates an important and oft-used concept in astrophysics, that of *dynamical time*, which is defined as

$$t_{\rm dyn} = T/4 = \sqrt{\frac{3\pi}{16 G_N \rho}},$$

which, of course, is also the answer to the question above, as the diligent Reader has verified by herself. The dynamical time for a system of average density ρ is thus, roughly, the time required for a test mass to travel halfway across the system.

The matter distribution for axisymmetric galaxies like our own Milky Way is evidently completely different from a homogeneous sphere. The calculation of the potential for cylindrically symmetric systems, for example disks, is significantly more complicated than for spherically symmetric ones, and it involves typically one of three possible approaches:

(i) elliptic integrals,
(ii) Bessel functions, or
(iii) a superposition of flattened spheroids (again, I refer the passionate Reader to Ref. [5]).

The gross features of the resulting circular velocities are, however, still a (1) close-to-linear rise for small r and (2) Keplerian fall-off at large radii.

Observations of rotation curves indeed showed a linear rise in the inner regions, followed by a radius at which $v_c(r)$ reached a maximum — the *"turnover point"*. Interestingly, though, most galaxies were not observed to feature the expected subsequent Keplerian fall-off — a fact that was initially simply attributed to the scarcity of statistics at large radii. After 1970, however, the improved observational capabilities of modern radio facilities allowed to conclusively assess that in fact *no Keplerian fall-off could be observed at all* in essentially any case, with $v_c(r)$ exhibiting a flat or very slowly declining large-r behavior. The simplest interpretation, as first put forth in 1970 by Freeman [6], is that there exists some "missing mass" in the form of a dark galactic halo surrounding spiral galaxies.

Assuming a spherical dark halo, flat rotation curves imply, at large radii, a linear increase of total halo mass with radius, and therefore

$$\rho(r) = \frac{1}{4\pi r^2}\frac{dM(r)}{dr} = \frac{v_c^2}{4\pi G_N r^2},$$

where $\rho(r)$ indicates the matter density as a function of Galacto-centric distance r, and where I am again assuming spherical symmetry. A common parameterization for ρ which accounts for the quasi-linear small r regime is the phenomenological fitting form (also known as isothermal sphere profile),

$$\rho(r) = \frac{\rho_0}{1+(r/a)^\gamma},$$

with $\gamma \simeq 2$ and a the characteristic linear scale of the system. Cosmological N-body simulations, possibly augmented with some attempt at modeling baryonic effects in addition to gravitationally interacting, dissipationless, collisionless dark matter particles, are at the moment the best tools to theoretically predict the shape of the dark matter halo density distribution profiles. I will discuss a few examples of dark matter density profiles in the next chapters; for more on the dark matter density and velocity distribution, see also Sec. 1.7.

An additional early "mechanical" argument for the existence of dark halos has to do with the stability of galactic disks. Numerical work on disk stability had shown since the early 1970s that self-gravitating disks tended to quickly form bars (*"bar instability"*), unless they were "hot", i.e. with very large velocity dispersion [7]. Ostriker and Peebles in 1973 famously confirmed this result, and found a quantitative criterion (the *Ostriker–Peebles criterion*) [8]: indicating with

$$t \equiv T/|U|,$$

the ratio of rotation kinetic energy to potential energy, and indicating with $\Pi/2$ the kinetic energy in random motion, a necessary condition for the stability of self-gravitating systems to bar-like modes was found to be

$$t < t_{\rm crit} = 0.14 \pm 0.02 \quad \text{or} \quad \Pi/T \gtrsim 5 \text{ (see Ref. [8])}.$$

The instability is cured if the disk is embedded in a dark halo whose mass inside the outer radius of the disk is comparable or greater than the disk mass. There exist, however, other ways to suppress disk instabilities [9], which essentially cut off the instability by dynamically shifting galactic mass to the inner regions, steepening the rotation curve near the galactic center. As a result, disk stability is not as compelling an argument to the necessity for the existence of dark matter in galaxies.

1.3 X-ray halos

Large astrophysical systems contain abundant baryonic[g] matter that does not shine (much). For example, interstellar gas. If the gas is "hot", as a result of sitting inside a deep gravitational potential well, it emits thermal bremsstrahlung photons at X-ray frequencies that can be detected.

> **Exercise 10.** Assuming virialization, estimate the temperature of gas for a cluster (typical size 1 Mpc, typical mass $10^{15}\ M_\odot$), for a Milky-Way-like galaxy (10 kpc, $10^{12}\ M_\odot$) and for a dwarf galaxy (1 kpc, $10^8\ M_\odot$).

As particle dark matter enthusiasts, we are interested in these X-ray halos for at least three reasons. First, they can give us handles to disentangle (mostly dark) gas from non-baryonic dark matter; Second, such X-ray halos are potential sources of background to particle dark matter detection (see e.g. Chapter 8). Third, maybe we can cook up models where dark matter particles interact with the dense thermal gas producing observable signals [10].

For a spherically symmetric system, Euler's equation of fluid dynamics (as is often the case, this is just a disguised $F = ma$),

$$\rho \frac{d\vec{v}}{dt} = -\nabla p - \rho \nabla \Phi \tag{1.6}$$

gives, for spherical systems in equilibrium (i.e. $\vec{v} = 0$),

$$\frac{dp}{dr} = -\frac{G_N M(r) \rho}{r^2}, \tag{1.7}$$

with ρ and p density and pressure, and $M(r)$ the mass within radius r.

---

[g]From the Greek word for "heavy", the term *baryons* indicates bound states of (strongly interacting) quarks, as opposed to leptons, from the Greek word for "light", indicating electrons, neutrinos, and their second and third generation relatives.

Exercise 11. Show that Eq. (1.6) follows from Eq. (1.7) for $\vec{v} = 0$ for spherically symmetric systems.

Using the ideal gas law

$$p = \frac{\rho k_B T}{m},$$

and assuming, for simplicity, that the gas is composed of protons only (of mass m_p), you get

$$M(r) = \frac{k_B\, T\, r}{G_N m_p}\left(-\frac{d\ln\rho}{d\ln r} - \frac{d\ln T}{d\ln r}\right).$$

This is great! Measuring $T(r)$ and the gas density profile $\rho(r)$, you can reconstruct the *total* mass density profile! Added benefits include that the gas pressure is isotropic, so there are no complications due to e.g. stellar orbits.

As you will show in the next exercise, X-rays are typically optically thin,[h] thus the *luminosity density* is just proportional to the square of the physical density. Now, with good-enough X-ray observations you can determine the temperature from the shape of the continuum X-ray spectrum, or from the strengths of X-ray lines,[i] and of course (by just counting photons) the gas density.

Exercise 12. Consider a mass M of gas uniformly distributed in a sphere of radius r_h, and calculate the number of scattering events for a photon traveling from the center out to r_h, i.e. $N \approx r_h n_{\rm gas} \sigma_T$, with σ_T the Thomson cross section. Plug in representative values for a Milky-Way-like galaxy, $r_h \sim 10$ kpc and and $M \sim 10^{11}\ M_\odot$.

This technique allowed for the early determination of the mass distribution around elliptical galaxies (in particular M87), and for the localization of collisional matter in systems like the Bullet cluster (also less poetically known as 1E 0657-558, Fig. 1.2; the red in the figure indicates where collisional matter, i.e. gas, is, as observed at X-ray frequencies; the blue regions

[h]This is a subtle but important point for indirect dark matter searches as well, so listen!

[i]Which come from de-excitation of mostly ionized $Z \sim 15$–20 atoms.

Figure 1.2: This is what you get if you google "dark matter".

are smeared-out mass density contours reconstructed via weak gravitational lensing: see Sec.1.4 for more on this).

1.4 Weak lensing

The observation that mass bends the trajectory of light is one of the many reasons why we take general relativity quite seriously (including models of the universe based on it!). Galaxies and clusters of galaxies act as gravitational lenses. One of the effects on bright, distant background sources (for example galaxies or quasars) is the possibility that intervening gravitational lenses produce multiple images of the same source (strong lensing) or distort the background source image (weak lensing). To see how this can be used to probe dark matter let us go through a few little exercises:

Exercise 13. Consider a non-relativistic test particle (say, a star) traveling with speed v past a spherically symmetric galaxy with a mass distribution $M(r)$. Suppose that the star's velocity is large enough for the deflection angle to be small, and call the distance of closest approach (a.k.a. impact parameter) b. Calculate $F_\perp$, and use $F = ma$ in

(Continued)

Exercise 13. (*Continued*)

its $m\dot{\vec{v}}_\perp = \vec{F}_\perp$ incarnation; integrate with respect to time, and show that the deflection angle is approximately

$$\alpha_{\text{non-rel}}(v) \simeq \frac{2G_N b}{v^2} \int_b^\infty \frac{dr}{\sqrt{r^2 - b^2}} \frac{M(r)}{v(r)}.$$

Exercise 14. Use the weak-field approximation to a FRW metric (see Appendix B), where ϕ is the deflecting (Newtonian) gravitational potential,

$$ds^2 = \left(1 + 2\frac{\phi}{c^2}\right) dt^2 - \left(1 - 2\frac{\phi}{c^2}\right) dl^2$$

and Fermat's principle to argue that:

(i) the deflection angle for light in the case of a point mass at impact parameter b is

$$\alpha = \frac{4G_N M}{c^2 b}.$$

(ii) $\alpha(v = c) = 2\alpha_{\text{non-rel}}(v)$ for a generic $M(r)$.

Exercise 15. Assume a galaxy has a flat rotation curve with circular speed v_c. Show that the deflection angle for photons is

$$\alpha = \frac{2\pi v_c^2}{c^2}.$$

Exercise 16. Show that the angular separation α of two source images s formed by a singular isothermal sphere ($\rho(r) \equiv \frac{\sigma^2}{2\pi G_N r^2}$, the lens l) is, in Euclidean space,

$$\Delta\theta \equiv \frac{2\alpha D_{sl}}{D_s} \equiv \frac{8}{\pi} D_{sl} \sigma^2 D_s c^2,$$

(*Continued*)

Exercise 16. (*Continued*)

where $D_{\rm sl}$ is the distance between the source and the lens (the center of the isothermal sphere), and D_s is the distance of the observer to the source.

I know you are now exhausted (or you should be), so please take a deep breath and stretch. Good news is what we wrapped our minds around above is a key method used in cosmology, and one that is sensitive to the *total* matter mass, thus directly probing the dark matter content of galaxies (see the blue mass contours in Fig. 1.2). One of the first examples came from observations of quasar Q0957 + 561, which had $D_{\rm sl}/D_s = 0.573$ (where, as in Exercise 16, $D_{\rm sl}$ is the source-lens distance, and D_s is the source distance) and an angular separation $\Delta\theta = 6.2''$.

Exercise 17. Calculate the rotation speed of the lensed galaxy for Q0957 + 561 assuming the rotation curve is flat.

In addition, we will use the results above to look for the first dark matter candidate we will talk about in this book, Massive Compact Halo Objects (MACHOs) (see Sec. 1.8).

Perhaps the most important impact of lensing studies on *particle* dark matter comes from probes of its collisional properties such as the "Bullet" cluster system 1E0657-56, shown in Fig. 1.2. There (and in other similar systems) lensing-reconstructed mass density profiles indicate two distinct massive sub-structures, with an off-set baryonic mass distribution traced by X-ray observations; clusters' galaxies are totally collisionless, and thus follow the dark matter density and not the thermal gas, challenging for example modified theories of gravity where the opposite should generically happen.

X-ray data also indicate a recent "collision" between the two galaxy clusters, and detailed hydrodynamical simulations allow to set limits on the self-interaction cross section of dark matter $\sigma_{\chi\chi}$ per unit (dark matter) mass m_χ on the order of [11]

$$\frac{\sigma_{\chi\chi}}{m_\chi} \lesssim 1\ {\rm cm}^2/{\rm g}. \tag{1.8}$$

This is actually a reasonable number (although, with somewhat weird units), as you will convince yourself carrying out the following exercise.

Exercise 18. Estimate the mean free path for dark matter self-interactions in a medium of (dark matter) density $\rho \sim 1$ GeV/cm^3 for $\frac{\sigma_{\chi\chi}}{m_\chi} \sim 1$ cm^2/g. Since the typical size of a cluster is on the order of 1 Mpc, argue that the limit in Eq. (1.8) makes sense.

Self-interactions can also affect the *shape* of halos, generically making halos rounder[j]; the resulting constraints are in the same range as those from Bullet-cluster-type constraints (similar densities and distance scales are involved).

Treating weak lensing on cosmological scale *statistically* is no easy feat, but is also something that has very interesting implications: it can be used to constrain the total non-relativistic matter density in the universe, as well as the fraction of *hot* dark matter in the universe, i.e. dark matter that was relativistic at the epoch when structures started to form in the universe. A great description of weak lensing in cosmology is given in Chapter 10 of Scott Dodelson's phenomenal cosmology textbook [12].

1.5 A simple subtraction

The SM matter density in the universe is essentially all in baryons.[k]

Exercise 19. Since the universe is charge-neutral, estimate the ratio of leptonic to baryonic density,

(i) neglecting neutrinos, and
(ii) including the maximal possible neutrino density allowed by cosmological observations (see e.g. Ref. [3] for the latter number, or skip to the end of this section).

Determining the density of any *non-baryonic* dark matter can thus be accomplished by (i) measuring the baryon density, (ii) measuring the total matter density, and (iii) subtracting the first from the second, a simple subtraction with a few, inevitable, subtleties.

[j]A caveat to this is the case of massless mediators of the dark matter self-interactions (I thank Jakub Scholtz for bringing this to my attention!).
[k]Hence the name!

There are four key ways to pinpoint the baryon density,

$$\Omega_b = \rho_b(\text{today})/\rho_{\text{crit}},$$

which are performed at rather different redshifts; taking the redshift into account is simple, since

$$\frac{\rho_b}{\rho_{\text{crit}}} = \Omega_b a^{-3}.$$

Most of the baryons in the universe are in the form of gas in groups of galaxies — and measuring, e.g. via X-rays, how much gas is in there yields $\Omega_b h^2 \sim 0.02$ (where the h is Hubble's constant in units of 100 km s^{-1} Mpc^{-1} — in practice $h^2 \simeq 0.5$, see Appendix B). Baryons get in the way of light emitted by distant sources, e.g. quasars. Observations of how light from quasars is absorbed depends on the amount ("column density") of baryons along the line of sight. While there is a sizable associated uncertainty, this also yields $\Omega_b h^2 \sim 0.02$. A third way is to use the spectrum of the CMB: the relative height of the odd and even peaks in the anisotropy power spectrum (and also the position of the peaks) depends sensitively on Ω_b. The first acoustic peak is strongly enhanced for increasing Ω_b, while the second peak is strongly suppressed, and both are shifted to higher l (smaller angular scales). This is a very fine probe of Ω_b. With Planck observations this gives [3] $\Omega_b h^2 = 02225 \pm 0.00023$ (with the assumptions specified for column 5 of Table 1 in Ref. [3]).

Finally, a really interesting test of the baryon abundance comes from comparing the predictions for the abundance of light elements synthesized at Big Bang Nucleosynthesis (BBN) (see the beautiful Chapter 3 of Ref. [12] or the classic Chapter 4 of Ref. [13] for details). Simply using measurements of the deuterium abundance one pinpoints $\Omega_b h^2 = 0.0205 \pm 0.0018$ [14]. The beauty of using BBN to predict the abundances of light elements is that this technique indirectly probes the validity of many of our extrapolations of general relativity, statistical mechanics and nuclear and particle physics we love to use to study the early universe, and it shows that it all beautifully and quantitatively works very well!

While lensing is one tool to constrain the total matter density in the universe (let us call it for brevity $\Omega_m = \rho_m/\rho_{\text{crit}}$ normalized to the critical density — see Appendix B), other probes exist. One consists of compiling measurements of mass-to-light ratios as a function of scale [15]: the ratio is observed to saturate at a roughly constant value past cluster-size scales (roughly 1 Mpc), implying $\Omega_m \sim 0.3$. Large-scale surveys can be used to quantify the power spectrum of the distribution of galaxies; such power spectrum depends sensitively on $\Omega_m h$, and data indicates that $\Omega_m h \sim 0.2$ [16]. The cosmic velocity field in relation to the distribution of galaxies also points to $\Omega_m \sim 0.3$ [17]. Finally, the amplitude of the power spectrum of anisotropies in the CMB at different angular scales also depends on $\Omega_m h^2$; CMB data indicate a value of $\Omega_m h^2$ consistent with other observations. The latest Planck results, in particular, give $\Omega_m = 0.308 \pm 0.012$ [3].

Somewhat less directly, one can use probes of the ratio Ω_b/Ω_m hoping to then use the rather accurate measurements of Ω_b to infer Ω_m. An example is the probe of the ratio of the mass of gas in clusters to the total mass, which can be performed by looking at X-ray emission or at the distortions to the CMB induced by the "heated" electrons in the cluster, and assuming that this ratio is a proxy for the baryonic-to-total matter density ratio in the universe as a whole. Other probes, which I will not review, but that I will only list here, are baryon acoustic oscillations and, again, CMB anisotropies.

The outcome of the "simple subtraction" is that the baryon density is roughly 5% of the critical density, and that the total matter density is roughly 30%. The difference, about a quarter of the universe critical density, is in the form of non-baryonic dark matter.

Note that neutrinos are in principle non-baryonic, and could contribute to some of this non-baryonic matter density. As we will discuss in what follows, this possibility is excluded by the type of structure formation history massive neutrinos would have produced in the early universe. In short, neutrinos would behave as "hot" dark matter — they would be relativistic at the epoch of structure formation, free-streaming over very large distance scales at the time matter structures start to gravitationally collapse and form halos. As a result, if neutrinos contribute a large-enough fraction of the inferred cosmological non-baryonic dark matter, they would suppress the formation of small-scale halos in a dramatic way, incompatible with observation. This can be turned into a quantitative limit on the relative abundance of neutrinos in the universe, $\Omega_\nu h^2 < 0.0025$ at 95% C.L. according to the most recent Planck results [3], which essentially makes them irrelevant to $\Omega_{\rm DM}$.

1.6 Timely structure formation

As a scholar of dark matter, I often get asked whether we are really sure that we need dark matter after all. Can it possibly be modified gravity after all? There was (justified) enthusiasm in popular science news after the discovery of the Bullet cluster system's X-ray versus gravitational potential map — baryons sit not where most of the gravitationally interacting matter, as traced by weak gravitational lensing, is (see Fig. 1.2). It is naturally very hard to explain observations of this type with modified gravity, even if one is clever enough to work out the inherently general-relativistic weak lensing formalism in modified theories of gravity (that have a hard time being formulated in a GR-friendly "covariant" way). However, my answer to the question "can it be modified gravity?" usually is along a different line of argument: given the observed "smoothness" of the CMB sky, it is almost impossible to explain the timely formation of non-linear structure in the universe unless baryons fall into pre-existing gravitational potential wells seeded by dark matter. The argument is quite simple, and can be easily made quantitative, as I review below.

The physical problem is simple: given an initial pattern of density perturbations, such pattern should be shared between the (tightly coupled) baryon and photon fluids in the early universe up to their mutual decoupling, which happens at "recombination" (of protons and electrons), or CMB decoupling. Interestingly, we can directly measure the size of these inhomogeneities by measuring angular anisotropies in the temperature fluctuations of the CMB sky (the baryonic density fluctuations $\delta\rho/\rho$ at recombination are a calculable multiple of the measured photon temperature fluctuations $\delta T/T$). These density fluctuations peak at about 1 part in 10^4, i.e. $\delta\rho/\rho \lesssim 10^{-4}$. General relativity dictates that for matter in the linear regime,[l] inhomogeneities grow as $\delta\rho/\rho \sim a$, a being the scale factor. In turn, the scale factor has grown, since recombination, roughly by a factor equal to the redshift at recombination, $z_{\rm rec} \sim 1,100$: If there are only baryons around, not enough time has passed for observed structures (which have $\delta\rho/\rho \gg 1$) to form!

CMB thus tells us that in a baryon-only universe there is not enough time for structures to get to the non-linear regime,[m] as illustrated by the dashed line in Fig. 1.3. Structures have time to go non-linear (the "1" line above) only if there is some other particle species that decoupled from the baryon-photon fluid *much earlier* than recombination, creating *seeds* of inhomogeneity (gravitational potential wells) much deeper than baryonic

[l]That is, $\delta\rho/\rho \lesssim 1$.

[m]"*A prediction that is embarrassingly wrong coming from a species that owes its existence to nonlinear structures.*" Scott Dodelson puts it in Ref. [18].

ones (i.e. $(\delta\rho/\rho)_{\rm DM} \gg 10^{-4}$ at recombination). In other words, for structures to form in time we need a dark matter species that is both sufficiently electrically neutral (dark) and non-relativistic enough (cold) to not, by itself, free-stream out of early gravitational potential wells. *Dark matter seeds timely structure formation*! There are a number of subtleties to the story I outlined above, but I believe that this general argument is perhaps the most compelling one to the existence of a dark matter in the universe.

Requiring timely structure formation directly rules out simple versions of modified theories of gravity [18]; in truth, certain TeVeS theories [20] do predict large enough inhomogeneities (see Fig. 1.3) for structures to form in time, i.e. hitting the "1" line, where $\delta\rho/\rho \sim 1$. However, such models vastly overpredict baryon acoustic oscillations (as they should) and the matter power spectrum they predict is typically in stark contradiction with data, at least to the level at which reliable cosmological predictions for such models can be made (compare the data points from Sloan [19] with the TeVeS blue solid line in Fig. 1.3, from Ref. [18]).

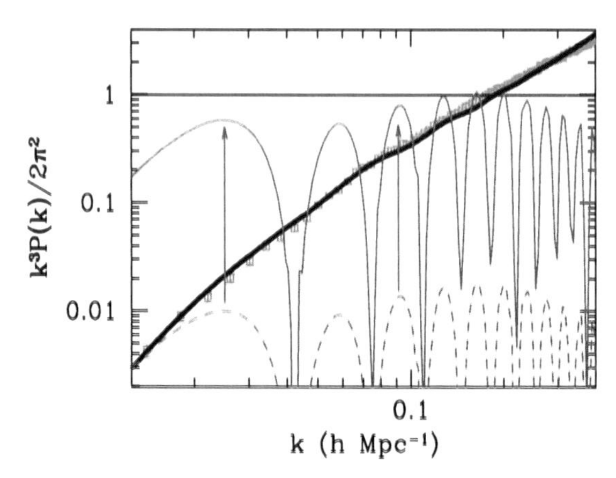

Figure 1.3: Modified gravity does not work. (The figure, adapted from Ref. [18], shows the power spectrum of matter density fluctuations; the points, with error bars, are from the Sloan Digital Sky Survey [19], the black line is the prediction for a cold dark matter plus cosmological constant cosmology model; the dashed line corresponds to a model without dark matter, with all matter density (20% of the critical density) in baryons. The solid line is an attempt at a covariant formulation of modified gravity (MOND) known as Tensor–Vector–Scalar model (TeVeS) [20]: the amplitude can reach unity, but the spectrum, with large baryon acoustic oscillations, is quite a bit different from data).

1.7 The dark matter density and velocity distribution

Classical mechanics teaches us many profound lessons. One of them is how phase-space (velocity, or momentum, and position, that is) density distributions, $f(\vec{x}, \vec{v}, t)$, evolve. In the absence of processes such as collisions that alter the phase-space density (in Chapter 3, we will call such processes with their name, and introduce the collision operator $C[f]$), in the non-relativistic, collisionless limit f obeys a simple conservation equation:

$$\frac{\partial f}{\partial t} + \dot{\vec{x}}\frac{\partial f}{\partial \vec{x}} + \dot{\vec{v}}\frac{\partial f}{\partial \vec{v}} = 0. \tag{1.9}$$

There is a general theorem, known as Jeans's theorem, that deals with steady-state ($\partial f/\partial t = 0$) solutions to the equation above: such solutions can only be functions of *integrals of motion*, i.e. functions $I(\vec{x}, \vec{v})$ that are conserved, $dI/dt = 0$. Conversely, any function of the integrals of motion is a steady-state solution to Eq. (1.9). Examples of possible integrals of motion, as you know from baby classical mechanics, include the Hamiltonian, angular momentum, a component of angular momentum, etc., depending on the symmetries of the problem. Spherically symmetric, Hamiltonian systems, in particular, have phase-space densities that can only depend on two integrals of motion: the total energy (Hamiltonian) and the magnitude of the (total) angular momentum, $f = f(E, L)$. If, additionally, a system has an isotropic velocity dispersion tensor $\langle v_i v_j \rangle$, then $f = f(E)$ [5].

Let's define a *relative energy*[n] $\mathcal{E} = \Psi - \frac{1}{2}v^2$, where Ψ is the gravitational potential, and consider a phase-space distribution $f(\mathcal{E}) \propto e^{\mathcal{E}}$. We can calculate the spatial density distribution from

$$\rho \propto \int_0^\infty dv\, v^2 f(v) = \int_0^\infty dv\, v^2 \exp\left(\frac{\Psi - v^2/2}{\sigma^2}\right) \propto e^{\Psi/\sigma^2},$$

where σ is the velocity dispersion: the equation above shows how, for spherical systems, there is a simple relation between the density and the velocity distribution. To take this a step further, use Poisson's equation

$$\nabla^2 \Psi = -4\pi G_N \rho$$

to get

$$\rho(r) = \frac{\sigma^2}{2\pi G_N r^2}.$$

[n] Energy per unit test particle mass.

We saw in Sec. 1.2 that measurements of the galactic rotation curves indicate that the dark matter density scales radially as $\rho(r) \sim 1/r^2$. Jeans's theorem then provides us with the associated velocity distribution,

$$f(v) \propto e^{-v^2/\sigma^2}. \tag{1.10}$$

Exercise 20. Derive Eq. (1.10).

Equation (1.10) is (proportional to) the phase-space distribution of a self-gravitating isothermal gas sphere (with $\sigma^2 = k_B T/m$, i.e. see [5] for more details). Of course, this simple picture is a bit too simple — for e.g. the total mass of the system, if described by the density profile above, diverges, rotations curves are not flat forever, etc.

One other significant shortcoming of the picture above is that the steady-state assumption is actually not that great: galaxies merge, and even today we see remnants of recent mergers in the form of (beautiful) stellar streams, which are certainly not virialized. This affects both the spatial density and the velocity distribution, with high-velocity tails. In addition, if structure formed hierarchically from the merger of smaller structures, some memory of the clumpy early epoch might remain in the form of clumps both in density and in velocity space [21]; other notable features include debris flow [22] from the aggregated effect of sub-halo mergers, and, possibly, a "dark disk" [23] of sub-halos preferentially disrupted and then dragged by the baryonic disk (thus co-rotating with the Sun). All of these effects add to the uncertainties in the description of the Galactic dark matter density and velocity distribution, whose determination, as mentioned, ultimately relies on results from N-body simulations and input from observations.

The reference velocity distribution for our galaxy the Dark Matter Scholar should bear in mind is inherited from the theoretical discussion above, and assumes a Maxwell–Boltzmann distribution, truncated at the galaxy's escape velocity,

$$f(\vec{v}) = \begin{cases} \frac{1}{N_{esc}} \left(\frac{3}{2\pi\sigma^2} \right)^{3/2} \exp\left(\frac{3\vec{v}^2}{2\sigma^2} \right), & |\vec{v}| < v_{esc} \\ 0 & |\vec{v}| \geq v_{esc}, \end{cases} \tag{1.11}$$

where $\sigma \simeq 288$ km/s [24] is the r.m.s. velocity dispersion, and $v_{esc} \simeq 544^{+64}_{-46}$ km/s is the escape velocity [25].

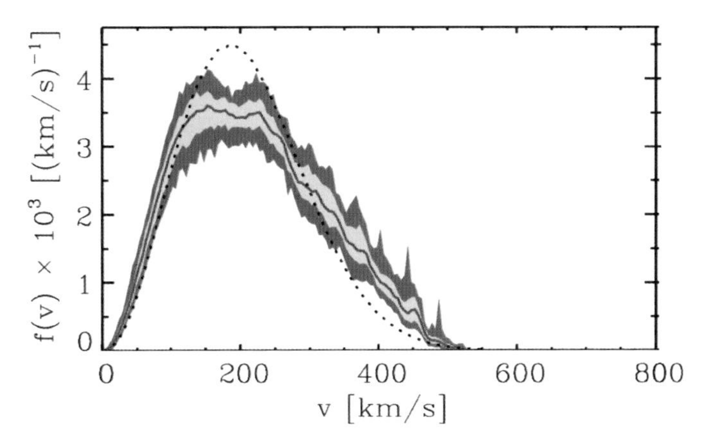

Figure 1.4: Not very smooth: The velocity distribution found with the *Via Lactea* simulation (solid red), with the 68% scatter (light green) and the maximum and minimum ranges (dark green), compared with the best-fit Maxwell–Boltzmann fit, from Ref. [26].

Exercise 21. Show that

$$N_{\rm esc} = {\rm erf}(z) - 2\pi^{-1/2} z e^{-z^2},$$

with $z \equiv v_{\rm esc}/(\sqrt{2/3}\sigma)$.

Figure 1.4, from my UCSC colleague Piero Madau and collaborators [26], compares the functional form of Eq. (1.11) with the results of their *via Lactea* (Milky Way in Latin) simulation. It is interesting to observe the high-velocity "tails" and "clumps" in velocity space.

Exercise 22. Estimate the escape velocity of the Milky Way, and compare with the number given above.

Exercise 23. Use the virial theorem to estimate the velocity dispersion of our Milky Way ($R_{\rm halo} \sim 100$ kpc, $M_{\rm halo} \sim 10^{12} M_\odot$), of a typical dwarf spheroidal galaxy satellite ($R_{\rm halo} \sim 10$ kpc, $M_{\rm halo} \sim 10^8 M_\odot$), and of a typical galaxy cluster ($R_{\rm halo} \sim 1000$ kpc, $M_{\rm halo} \sim 10^{15} M_\odot$).

1.8 Can "dark" baryons fit the bill?

In short, no, dark baryons cannot fit the bill. There are better and worse hidden baryons. The worse type is macroscopic objects that do not shine, also known as MACHOs, a much more marketable acronym, turns out, than that for a much "better" class of dark matter candidates, weakly interacting massive particles (WIMPs). We know that on cosmological scales baryons just do not work as dark matter: BBN limits the cosmological density of baryons to a fraction of the matter density, and it is hard to imagine that the missing baryons could pop out later; also, as Fig. 1.3 illustrates, structure formation really does not work with baryons only. However, people wanted to make sure that at the galactic scale (especially in the galaxy, i.e. our own Milky Way) the dark halo was not made up of Jupiter-like things (balls of gas just small enough to not end up forming stars), brown dwarfs, or any other macroscopic baryonic compact object.

The best way to search for MACHOs is not exactly very exciting: stare at as many stars in the Magellanic Clouds as possible, for as long as possible, and try to detect a small magnification, called *microlensing*, of the stars' brightness, caused by the passage of a compact object. The calculation of the gravitational amplification is a good little exercise which I invite you to carry out.

Exercise 24. Use the formula for the deflection of photons by a point-like massive body you derived above to show that the magnification factor for microlensing is

$$A = \frac{2+u^2}{u\sqrt{4+u^2}},$$

where $u = b/r_E$, and where $r_E = \sqrt{G_N M d}$ is the so-called Einstein radius, with $d = 4d_1 d_2/(d_1 + d_2)$ and d_i the distances between observer and lens and lens and star (the impact parameter b assumed to be much smaller than d_i). You can use Ref. [27] to guide your calculation.

Now we should estimate how long a microlensing event is expected to be, and how likely it is. There you go:

Exercise 25. We will estimate the duration of a microlensing event from a Milky Way MACHO as expected from observations of the Large Magellanic Cloud, about 50 kpc away.

(i) Calculate the Einstein radius for a MACHO of mass m and show that it is

$$r_E \simeq 1.2 \times 10^{12}\ \mathrm{m} \left(\frac{m}{M_\odot}\right)^{1/2} \left(\frac{\sqrt{d_1 d_2}}{25\ \mathrm{kpc}}\right).$$

(ii) Infer that for $m = M_\odot$ and for $d_1 = d_2 = 25$ kpc the time it takes for a MACHO to traverse the Einstein radius is about 2 months. This is a reasonable timescale for observations!

Exercise 26. We intend to estimate the frequency of MACHO microlensing events for observations towards the LMC.

(i) Assume the Milky Way has a flat rotation curve v_c, and infer the number density as function of radius if all of the density is in MACHOs of individual mass m.
(ii) Assume that the average velocity orthogonal to the line of sight equals v_c (argue that this is not unreasonable); show that the fraction of MACHOs that pass through the Einstein radius at a distance $r_\perp$ is independent of $r_\perp$ and equals $2\pi r_E v_c \Delta t$, where Δt is the MACHO crossing time.
(iii) Now integrate over r, the radial distance to the Galactic center, making sure to include the relative angle to the LMC, which is $\alpha \simeq 82^\circ$. Show (numerically, the integral is pretty nasty) that the resulting rate is on the order of

$$5 \times 10^{-6}\ \mathrm{year}^{-1} \left(\frac{M_\odot}{m}\right)^{1/2}.$$

The exercise above implies that if one monitors on the order of a million stars in the LMC for a year, one should see 5 microlensing events if dark matter is in the form of MACHOs. In actuality, for example, OGLE-III observations of the Small Magellanic Cloud included almost 6 million stars, and a total time of observation of 2870 days [28]. They ended up observing three candidate microlensing events [28]. The result indicates, for example, that MACHO's of a mass of 1 $M_\odot$ can at most provide a fraction $\lesssim$10% of the dark matter halo mass. Slightly more competitive results , especially at low masses, were obtained by the EROS collaboration [29]. Overall, the range of MACHO masses as 100% contributors to the Galactic dark matter ruled out by microlensing is vast (ranging from $0.6 \times 10^{-7} < M/M_\odot < 15$ [29]). It is fascinating to realize that searches for dark matter in the form of MACHOs actually helped develop the now flourishing science of extrasolar planets!

Perhaps the only[o] "particle" dark matter candidate that could be made of "ordinary" matter is a primordial black hole (PBH).[p] Producing black holes (BHs) in the early universe is easy. In fact, it is easy to the point that it sometimes becomes annoyingly problematic (too many PBH's are formed) unless the spectrum of primordial density fluctuation is scale-free (Harrison–Zeldovich spectrum), i.e.: $(\delta\rho/\rho) \sim M^{-\alpha}$ with $\alpha \simeq 0$. If the mass within the horizon exceeds the Chandrasekhar mass, a BH will form. In a "standard" cosmology it only takes density enhancements of a few percent to form BHs, pretty much at any time in the history of the universe. Such enhancements are possible in a variety of scenarios, including bubble collisions, certain inflationary scenarios, phase transitions,[q] etc. [32–34].

There are generally no firm predictions for the mass of PBH, meaning the PBH mass function is highly model dependent. However, light PBH would have evaporated by now by Hawking radiation emission. But how light is light?

---

[o]Notable exceptions to this statement are non-standard baryonic bound states, such as strangelets (which we will briefly review in Sec. 9.3) and exotica such as Λ di-baryons [30].

[p]There are obviously not enough stars in the universe to produce a large enough stellar BHs density to explain the dark matter, hence the only option is to have BHs be produced in the early universe — hence the name "primordial".

[q]For example, first-order phase transitions have a "soft" equation of state, $p < \rho/3$, so if pressure support is weaker objects collapse more efficiently; this is possibly the case e.g. at the most recent phase transition we know of, the quark-hadron quantum chromodynamics (QCD) phase transition at $T \approx 100$ MeV [31].

Exercise 27. The Hawking temperature of a Schwarzschild (i.e. non-rotating, uncharged) BH of mass m is, in natural units,

$$T = \frac{1}{8\pi m}.$$

The Schwarzschild radius is $r_S = 2G_N m$. Hawking radiation for a BH of area $A = 4\pi r_S^2$ has a power

$$P = \sigma A T^4.$$

Calculate $m(t)$ and find the mass such that $m(t = t_U) = 0$, with $t_U \sim 10\,\mathrm{Gyr}$.

Exercise 27 should have convinced you that the magic number is roughly 10^{15} g, or $5 \times 10^{-19}\, M_{\odot}$, or 6×10^{38} GeV. Masses in the range $10^{17} \lesssim m/\mathrm{g} \lesssim 10^{20}$ would create an interesting interferometry pattern in the spectrum of gamma-ray bursts called femtolensing [35].[r]

Exercise 28. Verify that indeed for BHs in the mass range $10^{17} \lesssim m/\mathrm{g} \lesssim 10^{20}$ the Schwarzschild radius corresponds to the wavelength of gamma-ray photons.

Masses in excess of 10^{25} g are, instead, constrained by microlensing observations. The range in between can be probed by better femtolensing data, or distortions of the CMB from the non-observation of effects of accretion onto PBH [36].[s] However another possibility to constrain PBHs exists.

[r]This effect relies on the fact that the photon wavelength of gamma rays is comparable to the Schwarzschild radius for the BHs in the mass range, as the motivated Reader can verify (Exercise 28).

[s]After the epochal discovery of gravitational radiation by the LIGO collaboration [37], Ref. [38] suggested that BHs in the mass range inferred from the first LIGO event might be the dark matter; while this is in principle a possibility, the mentioned CMB constraints (see e.g. Ref. [36]), albeit somewhat systematically uncertain, could exclude this possibility.

If PBHs are gravitationally captured by neutron stars, at least for certain equations of state for the nuclear matter, they rapidly accrete and disrupt the star. The existence of old neutron stars in regions with large dark matter density suggests that PBH are a sub-dominant component of the dark matter across the whole mass range between 10^{15} and 10^{24} g, where microlensing constraints set similarly strong constraints. Although caveats can be put forward, and the PBH mass region between 10^{17} and 10^{24} g might still be in principle viable, there is in my opinion more than good reason to contemplate non-SM candidates for dark matter.

1.9 Gross features of a decent dark matter particle

While we have a rather accurate notion of *how much* dark matter there is on cosmological scales, and a pretty good amount of information on how dark matter is distributed on a broad range of scales, from dwarf galaxies ($M \sim 10^8\, M_\odot$) to the scale of clusters of galaxies ($M \sim 10^{15}\, M_\odot$), little do we know about exquisitely *particle* properties of the dark matter.[t] This section provides a quick summary of "gross" guidelines for particle dark matter model building.

Dark. Ahem, dark matter first off should not shine. However, how this translates *quantitatively* into constraints on the dark matter electric charge (or effective fractional charge, or "milli-charge" [39]) and on its electric or magnetic dipole moment [10, 40] is rather model-dependent, and mass-dependent (it is easier to hide a charge if the dark matter particle is ultra-heavy: what is effectively constrained is a ratio of charge to some power of the mass). One important piece of phenomenology associated to the dark matter being optically dark is that the process of loosing energy via photon radiation must be quite inefficient for dark matter (i.e. the dark matter is effectively *dissipationless*). What this implies is that dark matter particles would generally not accrete on BHs or collapse at the center of galaxies as efficiently as baryons (which cool electromagnetically). An interesting exception is a non-trivial dark sector that includes some light "dark photon" that would indeed help the dark matter dissipate and thus cool [41], and perhaps form "dark disks" [42].

[t]Of course, both the cosmological abundance and the density distribution of dark matter can be used to derive constraints on particle dark matter models, as we will see repeatedly in this book.

Collisionless (*really?*). We noted above that the dark matter self-interaction to mass ratio, $\sigma_{\chi\chi}/m_\chi$ is constrained by observations of cluster mergers and of the ellipticity of galactic halos. The dark matter is a *collisionless* particle. But is it, really? In practice we noted that the figure of merit for such constraints is given by a mean free path smaller than galaxy clusters scales (Mpc) for typical galaxy cluster dark matter densities (say, $1\,\mathrm{GeV/cm^3}$). This gives, for a dark matter mass of order the proton mass, an interaction cross section which can be as large as that of strong interactions (of order a barn, or the proton self-interaction cross section). I leave it to the Reader to decide whether or not this is a small cross section (not).

Classical. There exist some interesting constraints on the quantum nature of the dark matter particle not "interfering" with observation. For example, dark matter is observed to be confined on galactic scales as small as a kpc (dwarf galaxies). As a result, the de Broglie wavelength of the dark matter particle *must* be smaller than that, to have a "coherent" (in the quantum sense) dark matter halo. The following exercise quantifies this statement.

Exercise 29. Show that the de Broglie wavelength for a dark matter particle of mass m with a velocity typical of galaxies, say 100 km/s, is

$$\lambda \sim 3\,\mathrm{mm}\left(\frac{1\,\mathrm{eV}}{m}\right).$$

Show that for that dark matter particle to be confined on kpc scales, its mass must be larger than 10^{-22} eV.

Constraints are much stronger if the particle candidate is a *fermion*: in this case, Pauli blocking limits the phase-space density to, at most, $f = gh^{-3}$, with g the number of internal degrees of freedom.

Exercise 30. A typical galactic halo has a Maxwellian velocity distribution with $\sigma \sim 150$ km/s, and density $\rho \gtrsim 1\,\mathrm{GeV/cm^3}$. Argue that because of the exclusion principle a spin 1/2 fermion dark matter candidate must thus have mass

$$m^4 > \frac{\rho h^3}{[g(2\pi\sigma^2)^{3/2}]} \sim (25\,\mathrm{eV})^4.$$

This idea was first exploited by Tremaine and Gunn in their classic 1979 paper [43]. We will get back to the Tremaine–Gunn limit on several occasions later in the book.

Fluid. For "macroscopic" dark matter particles, say point masses (BHs?) of mass much bigger than the solar mass $M_\odot$, one might expect effects on the stability of bound systems. For e.g. large dark matter particles populating the dark halo of disk galaxies might disrupt the disk (by heating it). Ref. [44] calculates that this happens for $m \gtrsim 10^6\, M_\odot$. Massive dark matter projectiles would also tidally disrupt quiet balls of stars such as globular clusters, which hang around in the dark halo. Let's stick some numbers in for this case: it is instructive because a similar calculation concerns how stellar encounters can tidally disrupt mini-halos of dark matter (with stars and dark matter playing inverted roles). See e.g. [45].

Exercise 31. Consider a BH of mass $M_{\rm BH}$ and asymptotic velocity $V_{\rm BH}$ flying at an impact parameter b off of a globular cluster of mean square radius $\bar{r}$ and mass $M_{\rm GC}$.

(i) Show that the typical change in velocity of a star in the cluster is, in the impulse approximation (i.e. when the effective encounter time is much smaller than the typical crossing time of the system)

$$\delta V = \frac{2G_N M_{\rm BH}}{bV_{\rm BH}}.$$

(ii) Show that if $b^2 \gg \bar{r}^2$ the encounter causes a net increase in the kinetic energy of the globular cluster

$$\Delta E_{\rm far} = \frac{4G_N^2 M_{\rm BH^2} M_{\rm GC} \bar{r}^2}{3V_{\rm BH}^2 b^4}.$$

(iii) Argue that for the opposite limit, $b = 0$, the total change in kinetic energy can be cast as

$$\Delta E_{\rm near} = \pi \int_0^\infty [\delta V(R)]^2 \Sigma(R) R dR,$$

with $\Sigma(R)$ the globular cluster's surface density (see Ref. [5] for more details on this).

(iv) Substitute for $\delta V(R)$ in the expression above, and find an adequate minimal impact parameter to regularize the integral at small R.

(*Continued*)

Exercise 31. (*Continued*)

(v) The number of encounters can be quantified with a model for the dark matter halo density and velocity distribution; a limit on M_{BH} can then be set by requiring that globular clusters survive long enough (they are observed after all!), i.e. simplistically, that the total energy transfer over the age of a globular cluster (or of the galaxy) be equal or smaller than the globular cluster's binding energy,

$$E_{\rm bind} = -kG_N M_{\rm GC}^2/\bar{r},$$

with k a constant of order 1 that depends on the concentration of the cluster.

(vi) Estimate such limit!

Carrying out Exercise 31 in detail would reveal (under some assumptions) that a macroscopic dark matter particle of mass greater than $10^3\ M_\odot$ would disrupt globular clusters. Other constraints on the "granularity" of dark matter exist from Lyman-α observations which probe the power spectrum of density fluctuations, but they are generally weaker ($m \gtrsim 10^4\ M_\odot$) [46]. Additionally, as discussed above microlensing limits the fraction of dark matter in objects of macroscopic masses for a broad range of masses. Yet, since $10^3\ M_\odot \sim 10^{70}$ eV the range of possible particle dark matter masses is very, very large.

In conclusion, the *"name of the game"* for particle dark matter model building is schematically defined by:

(i) **Mass:** almost complete freedom for the particle mass (roughly 90 orders of magnitude available for bosons, 70 for fermions!);

(ii) **Interactions:** relatively weak constraints on interaction properties (both of the dark matter with itself — interactions that can be as large as strong interactions — and of the dark matter with ordinary matter, outside special mass ranges, see Chapter 4); but

(iii) **Abundance and Distribution:** a very good idea of the total (cosmological) dark matter density, and a pretty good idea of the density and velocity distribution at Galactic scales and beyond (not much as good information below those scales, though).

Time to arm ourselves with a good dose of ingenuity and creativity and to come up with ideas about what the particle making up most of the matter of the universe is. As Dr. Seuss says: *"Think left and think right and think low and think high. Oh the things you can think up, if only you try!"*.

1.10 Outline of the book

I organized the material in this book by first presenting some classic, "zeroth order" lessons on particle dark matter from cosmology (Chapter 2): the discussion guides you through estimates of the thermal relic density of hot and cold relics from the early universe, and of the mass range that thermal relics are expected to populate.

The following Chapter 3 puts the discussion on more solid grounds, and presents important caveats to the simple story of Chapter 2. Also, I present a few variations on the theme of thermal relics worth keeping in mind, and fun to work through.

Chapters 4–6 discuss ways in which we think we can get ahold of dark matter as a particle: *direct* detection (Chapter 4), *indirect*, or astrophysical detection (Chapter 5), and searches with particle colliders (Chapter 6). The implicit working assumption in these chapters is that dark matter is something resembling a WIMP, although often the discussion applies more broadly.

While this book is not a review, I devote three chapters to discussing a few specific particle dark matter candidates, in particular axions (Chapter 7) and sterile neutrinos (Chapter 8): both classes of models are interesting for reasons that go beyond the question of dark matter, and the detection methods and implications for cosmology and astrophysics are unique and important to be aware of. The final Chapter 9 is what I called a "bestiarium[u]": a collection of particle dark matter models that for a variety of reasons I consider especially compelling, including because they offer the opportunity to learn some beautiful nuggets of physics.

Enjoy!

[u]Medieval compendia of animals, where the history and illustration of each beast was accompanied by a moral lesson, reflecting the belief that every living thing had its own special meaning.

Chapter 2

The Thermal Relic Paradigm: Zeroth-Order Lessons from Cosmology

2.1 Praeludium

One of the most important lessons from cosmic microwave background (CMB) data is that the universe is nearly flat, and it thus has a density very close to the critical density (by definition of critical density; see Appendix B). It is therefore a good idea to keep in mind what the critical density is:

$$\rho_{\rm crit} \equiv \frac{3H_0^2}{8\pi G_N} \simeq 10^{-29}\ {\rm g/cm}^3,$$

where H_0 is the Hubble expansion rate today, and G_N is as usual Newton's gravitational constant. I hope the Reader appreciates how small the critical density actually is.

Exercise 32. How many hydrogen atoms are there in one cubic meter in the universe on average? What is the typical over-density in Galaxies like the Milky Way (of radius, say, 10 kpc, and mass, say, $10^{12}\ M_\odot$) compared to the average energy density of the universe? What about a cluster of galaxies (of radius, say, 1 Mpc and mass, say, $10^{15}\ M_\odot$)?

Our business here is to study dark matter, whose average density in the universe is

$$\bar\rho_{\rm DM} = \Omega_{\rm DM}\rho_{\rm crit} \simeq 0.3\rho_{\rm crit}.$$

It is useful to have on the tip of your tongue the value of this latter quantity to not only in the "cgs" units above, but also both in "astronomical"[a] and "particle physics" units:

$$\rho_{\rm crit} \simeq 3 \times 10^{10} \, \frac{M_\odot}{\rm Mpc^3} \simeq 10^{-6} \, \frac{\rm GeV}{\rm cm^3}.$$

Using "astronomical" units you could have immediately figured out the answer to Exercise 32: clusters of galaxies, the largest bound dark matter structures in the universe, have typical over-densities of 10^5 (indeed structures are highly non-linear, $\delta\rho/\rho \gg 1$!), since they host hundreds to thousands of galaxies, whose mass is in the $\sim 10^{12}\, M_\odot$ range; from the "particle physics" units, we learn, instead, that in our particular location in the Milky Way, where $\rho_{\rm DM} \sim 0.3\,{\rm GeV/cm^3}$ (another very important little number to keep in mind), the over-density is a factor of a few larger than in a typical cluster.

2.2 Thermal relics

A successful framework for the origin of species in the early universe is the paradigm of *thermal decoupling*. Thermal decoupling is an example of a fascinating cross-disciplinary synergy: it uses elements of *statistical mechanics*, of *general relativity*, and of *nuclear and particle physics* to make predictions that are often tested to exquisite accuracy with *astronomical* observations!

Broadly intended, the thermal relic framework encompasses the successful paradigms and predictions of recombination (i.e. the production of the CMB) and of Big Bang nucleosynthesis (BBN) (the production of light elements in the early universe), and it describes in detail, and as far as we know successfully, the process of cosmological neutrino decoupling. In short, thermal decoupling consists of the process where a particle species (photons, nuclei, etc.), at high temperatures, participates in a reaction that keeps it in "thermal equilibrium" in the statistical mechanical sense; as the universe cools, the particles' number densities decrease, and eventually

[a]Particle physicist: it is *always* a good idea to talk to astronomers; for example, I found my wife that way!

the *rate* Γ for that reaction becomes smaller than the expansion rate of the universe at that temperature, $H(T)$. Since the inverse of the latter (the "Hubble time" $t_H = 1/H(T)$) is a measure of the age of the universe at that point in time/temperature, and the inverse reaction rate is the measure of how long it takes for the reaction to occur on average, $\Gamma \ll H(T)$ indicates that *the reaction keeping the particle species in the universe is too slow*: on average, a particle undergoes less than one reaction over the age of the universe. The particle species is no longer in thermal equilibrium. This is called *thermal decoupling*, and the species is said to have *frozen out* of thermal equilibrium.

The term "freeze-out" indicates the moment, or the temperature $T_{\text{f.o.}}$, when the interaction rate $\Gamma(T_{\text{t.o.}})$ that keeps the species of interest in equilibrium becomes of the same order of the Hubble expansion rate $\Gamma(T_{\text{t.o.}}) \sim H(T_{\text{t.o.}})$ (see the dedicated Appendix B.3 for a detailed derivation of this criterion); after this point in time/temperature the particle species simply "redshifts" its number density away: Calculating the number density at any subsequent time only requires accounting for the expansion of the universe.

By definition of cross section (see Appendix A.2), the particle interaction rate is the product

$$\Gamma = n \cdot \sigma \cdot v$$

of a "target" particle number density n, the interaction cross section σ, and the "beam-target" relative velocity v.

Exercise 33. Make sure Γ as defined above has the dimensions of a rate (inverse time).

We will argue below that for several cases of interest here $v \sim c$, up to a factor of order 3–4, even for cold relics, and since in this chapter we are only interested in order-of-magnitude calculations, it will be fine to take $v \sim c$ for now; let us remember to carry out the cross check later, though!

Statistical mechanics teaches us that the equilibrium number density of a particle of mass m in a thermal bath of temperature T has two asymptotic regimes (see also Appendix B; note that I am assuming the species has zero chemical potential)[b]:

$$n_{\rm rel} \sim T^3 \quad \text{for } m \ll T,$$
$$n_{\rm non\text{-}rel} \sim (mT)^{3/2} \exp\left(-\frac{m}{T}\right) \quad \text{for } m \gg T.$$

The right-hand side of $\Gamma \sim H$, i.e. $H(T)$, comes from general relativity, and specifically from Friedmann's equation:

$$H^2 = \frac{8\pi G_N}{3}\rho.$$

In the radiation dominated epoch (i.e. $T \gtrsim 1$ eV),

$$\rho \simeq \rho_{\rm rad} = \frac{\pi^2}{30} \cdot g \cdot T^4,$$

with g the number of relativistic degrees of freedom ($g = 2$ for photons). To a decent degree of approximation, and recalling that the *reduced* Planck mass $M_P = 1/\sqrt{8\pi G_N} \simeq 2.435 \times 10^{18}$ GeV, in order to eyeball when thermal decoupling occurs you can use $H \simeq T^2/M_P$.

2.3 Hot and cold thermal relics

Let us put all of this in practice with a first simple example, and estimate the freeze-out temperature for Standard Model (SM) neutrinos. The relevant processes that keep the neutrino number density in equilibrium in the early universe are, schematically,

$$\nu + \bar{\nu} \leftrightarrow f + \bar{f},$$

with f a light charged fermion (the bar indicates antiparticles). We estimate the relevant scattering cross section for processes like that in the Fermi four-fermion contact interaction approximation (Appendix A.3), and we take $E \sim T_\nu$, so that $\sigma \sim G_F^2 T_\nu^2$ (where $G_F \sim 10^{-5}$ GeV^{-2} is Fermi's constant[c]).

[b]If a species' number density in the early universe is much smaller than that of photons, whose chemical potential is zero, then naturally $\mu \simeq 0$, as pointed out e.g. in Weinberg's classic book [47]. Convince yourself of the validity of this statement (Exercise 34).

[c]Note that we are implicitly assuming that the effective theory is valid, in this case that $E \sim T_\nu \ll m_W$! We will shortly cross check the validity of this assumption after calculating the freeze-out temperature!

At the neutrino freeze-out temperature $T = T_\nu$ we thus demand

$$n(T_\nu) \cdot \sigma(T_\nu) = H(T_\nu) \quad \rightarrow \quad T_\nu^3 G_F^2 T_\nu^2 = T_\nu^2 / M_P,$$

and therefore

$$T_\nu = (G_F^2 M_P)^{-1/3} \simeq (10^{-10} \times 10^{18})^{-1/3}\,\text{GeV} \sim 1\,\text{MeV}.$$

Among the various things to be happy about, we cheerfully verify that the T_ν we found indeed satisfies $T_\nu \gg m_\nu$, which we had implicitly assumed for the form of $n(T)$; we also learn that neutrinos are *hot relics*, not because they are particularly attractive, but because they freeze out while they are relativistic (vice versa a cold relic, to be discussed below, is one where the freeze-out happens in the $T_{\text{freeze-out}} \ll m$ regime, with m the relic's mass). Also, we are happy because the regime where we are using Fermi's effective theory of weak interaction is fine: $T_\nu \ll m_W$.

Let us now calculate the actual relic density for this "hot" thermal relic. Let me introduce the notation $Y = n/s$ where n is a number density and s is the entropy density. In an iso-entropic universe (one where entropy is conserved), $s \cdot a^3 = \text{constant}$, where a is the universe's scale factor. $Y \sim na^3$ is thus proportional to a "co-moving" number density. If no entropy is injected in the universe between the freeze-out temperature T_ν and T_{today}, then $Y_{\text{today}} = Y_{\text{freeze-out}} = Y(T_\nu)$: in other words, the only thing you need to worry about after freeze-out is the expansion of the universe.

In the case of hot relics, like SM neutrinos, keeping in mind that

$$Y_{\text{freeze-out}} = \frac{n(T_\nu)}{s(T_\nu)} = \frac{\rho_\nu(T_\nu)}{m_\nu \cdot s(T_\nu)}$$

and that, if entropy is conserved,

$$n_{\text{today}} = s_{\text{today}} \times Y_{\text{today}} = s_{\text{today}} \times Y_{\text{freeze-out}} \quad \text{(iso-entropic universe)},$$

the *physical density* in the late universe $\rho_{\nu,\text{today}} = m_\nu \cdot n_{\text{today}}$ is just

$$\rho_{\nu,\text{today}} = m_\nu \times Y_{\text{freeze-out}} \times s_{\text{today}} \quad \text{(iso-entropic universe)}.$$

For example, the fraction of the universe's critical density times h^2 (where h is today's Hubble constant in units of 100 km/s/Mpc — in practice

$h^2 \simeq 0.5$) in a SM neutrino species is

$$\Omega_\nu h^2 = \frac{\rho_\nu}{\rho_{\rm crit}} h^2 \simeq \frac{m_\nu}{91.5\ {\rm eV}}.$$

While the normalization depends on the relevant cross section (through the temperature dependence of the relativistic degrees of freedom at freeze-out,[d]) it is a general fact that *a hot relic's thermal relic abundance scales linearly with the relic's mass*. For a weakly interacting dark matter particle, requiring that the thermal dark matter density be less than or equal to the observed matter density, $\Omega_m h^2 \lesssim 0.13$ leads to the updated version of the so-called Cowsik–McClelland limit [48] on the mass $m_{\rm hot}$ of a hot dark matter relic — $m_{\rm hot} \ll 10$ eV. In practice, however, much better constraints on the mass of thermally produced hot relics come from structure formation [3].

Now let us worry about the opposite regime, the one where the relevant number density is in the *non-relativistic* regime (and again bear with me in neglecting v for now, we shall take care good care of velocities in the next chapter). The following exercise will make you realize that in a very important case we need the cold relic calculation[e]: how many baryons and antibaryons are left over from annihilation in the early universe?

Exercise 35. Calculate the freeze-out temperature $T_{\rm f.o.}$ for the $p\bar{p}$ annihilation reaction (you can use for example $\sigma \sim m_\pi^{-2}$ or $\sigma \sim \Lambda_{\rm QCD}^{-2}$) and estimate the relic proton/antiproton density; is this a hot relic problem? Compare what you find with the observed "baryon asymmetry".

A *cold relic* is one for which the freeze-out temperature is much lower than the mass of the particle (hereafter indicated with m_χ), which thus decouples in the non-relativistic regime, $T/m_\chi \ll 1$. An example of a cold relic is a "heavy" fourth neutrino N, with a mass $m_N \gg 1$ MeV.[f]

---

[d] For a species that decouples when relativistic, $Y_{\rm freeze\text{-}out} \simeq 0.278 \frac{g_{\rm eff}}{g_{*s}(T_{\rm freeze\text{-}out})}$, see also Appendix B and Chapter 5 of Ref. [13].

[e] And also why we need a mechanism for *baryogenesis*!

[f] As we will see in a couple of chapters, this heavy neutrino does not work as a thermal relic dark matter candidate — we would have seen such heavy neutrinos in direct detection experiments a long time ago!

In the non-relativistic "cold" case, the appropriate asymptotic form for the equilibrium number density is the non-relativistic limit

$$n \sim (m_\chi T)^{3/2} \exp\left(-\frac{m_\chi}{T}\right).$$

The freeze-out condition $n \cdot \sigma \sim H$ yields

$$n_{\rm f.o.} \sim \frac{T^2_{\rm f.o.}}{M_P \cdot \sigma}. \tag{2.1}$$

Let me call $m_\chi/T \equiv x$; using x, the cold relic regime corresponds to $x \gg 1$. We can then recast the condition $n \cdot \sigma \sim H$ as

$$\frac{m_\chi^3}{x^{3/2}} e^{-x} = \frac{m_\chi^2}{x^2 \cdot M_P \cdot \sigma}.$$

We thus need to solve

$$\sqrt{x} \cdot e^{-x} = \frac{1}{m_\chi \cdot M_P \cdot \sigma}, \tag{2.2}$$

with the added cold-relic requirement that $x \gg 1$. The function $\sqrt{x} \cdot e^{-x}$ is shown in Fig. 2.1, together with a few horizontal lines to guide the eye. What should we expect for the right-hand side in Eq. (2.2)? Let us, for example, substitute values we would expect for an "electro-weak interacting" cold relic (what we called a "weakly interacting massive particle (WIMP)"): being cold, we will use $E \sim m_\chi$, and cross fingers that Fermi's four-point

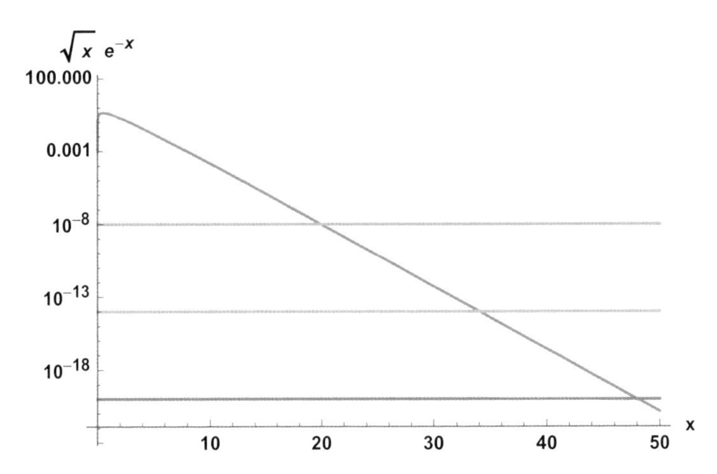

Figure 2.1: The function $\sqrt{x} \cdot e^{-x}$, in blue, and a few horizontal lines to guide the eye at 10^{-8} (yellow), 10^{-14} (the "WIMPs" value, green), and 10^{-20} (red).

weak interactions effective theory makes sense: let us plug in $\sigma \sim G_F^2 m_\chi^2$ and $m_\chi \sim 10^2$ GeV.

$$\sqrt{x} \cdot e^{-x} = \frac{1}{m_\chi \cdot M_P \cdot \sigma} \sim \frac{1}{10^2 \cdot 10^{18} \cdot 10^{-6}} \sim 10^{-14}.$$

This is the WIMP-y green line in Fig. 2.1, indicating $x \sim 35$. Suppose instead you had a hidden-sector mediator of mass $m_\phi \sim 1$ MeV, and a dark matter particle of mass $m_\chi = 1$ GeV, with an interaction cross section for the annihilation to SM particles $\sigma \sim \alpha^2/m_\phi^2$, and $\alpha \sim 10^{-2}$. Then you would have to solve $\sqrt{x} \cdot e^{-x} \sim 10^{-20}$ (red line in Fig. 2.1). Same thing but with $m_\phi \sim 100$ GeV and $\alpha \sim 10^{-3}$ it would give $\sqrt{x} \cdot e^{-x} \sim 10^{-8}$ (yellow line in Fig. 2.1).

As you can see from the figure, numerically the (generous) range $10^{-8} \ldots 10^{-20}$ for the right-hand side of Eq. (2.2) gives a range for $x_{\rm f.o.} \simeq 20 \ldots 50$. Now,

$$\Omega_\chi = \frac{m_\chi \cdot n_\chi(T = T_0)}{\rho_c} = \frac{m_\chi\, T_0^3}{\rho_c}\frac{n_0}{T_0^3},$$

with $T_0 = 2.75$ K $\sim 10^{-4}$ eV. Since for an iso-entropic universe $aT \sim$ const,

$$\frac{n_0}{T_0^3} \simeq \frac{n_{\rm f.o.}}{T_{\rm f.o.}^3},$$

and we have

$$\Omega_\chi = \frac{m_\chi\, T_0^3}{\rho_c}\frac{n_{\rm f.o.}}{T_{\rm f.o.}^3} = \frac{T_0^3}{\rho_c} x_{\rm f.o.} \left(\frac{n_{\rm f.o.}}{T_{\rm f.o.}^2}\right) = \left(\frac{T_0^3}{\rho_c M_P}\right)\frac{x_{\rm f.o.}}{\sigma}, \tag{2.3}$$

where I used Eq. (2.1) in the last step. The equation above can then be cast, plugging in the numbers for the various constants, as

$$\left(\frac{\Omega_\chi}{0.2}\right) \simeq \frac{x_{\rm f.o.}}{20}\left(\frac{10^{-8}\ {\rm GeV}^{-2}}{\sigma}\right). \tag{2.4}$$

Exercise 36. Use the equipartition theorem, in its natural-units incarnation

$$\frac{3}{2}T = \frac{1}{2}mv^2,$$

to derive a range for v at freeze-out if $x_{\rm f.o.} \sim 20 \cdots 50$.

Often, Eq. (2.4) is quoted with the thermally averaged product of the cross section times velocity $\langle\sigma v\rangle$ (we will understand why this is, and what a thermal average is, in the next chapter), instead of the simple cross section σ. Since, as you calculated in Exercise 36 $v \sim c/3$ for $x \sim 20$, one has

$$\langle\sigma v\rangle \sim 10^{-8}\ \mathrm{GeV}^{-2}\left(3\times 10^{-28}\ \mathrm{GeV}^2\ \mathrm{cm}^2\right)\ 10^{10}\ \frac{\mathrm{cm}}{\mathrm{s}} = 3\times 10^{-26}\ \frac{\mathrm{cm}^3}{\mathrm{s}}.$$

$\langle\sigma v\rangle \sim 3\times 10^{-26}\ \mathrm{cm}^3/\mathrm{s}$ is a "miracle" number definitely worth keeping in mind!

The "miracle" cross section for cold relics is sometimes associated with electroweak-interacting and electroweak-mass scale dark matter candidates, $m \sim E_{\rm EW} \sim 200$ GeV, under the header "WIMP miracle". The reason is simple: estimating the electroweak pair-annihilation cross section at $E \simeq T_{\rm f.o.}$ as

$$\sigma_{\rm EW} \sim G_F^2 T_{\rm f.o.}^2 \sim G_F^2\left(\frac{E_{\rm EW}}{20}\right)^2 \sim 10^{-8}\ \mathrm{GeV}^{-2},$$

one "miraculously" gets a cold thermal relic with $\Omega_\chi \sim 0.2$! Additionally, the electroweak scale is a plausible place to expect new physics based, for example, on naturalness arguments such as the hierarchy problem.

Is the "magic number" we just found unique and peculiar to the electroweak scale? Not at all! Let us remind ourselves which ingredients we used to get the "right" relic density:

(i) the condition for having a cold relic, $x \gg 1$, which via Eq. (2.2) translates into $m_\chi \cdot \sigma \cdot M_P \gg 1$,
(ii) a cross section $\sigma \sim 10^{-8}\ \mathrm{GeV}^{-2}$.

Now, suppose that the cross section be from dimensional analysis and given the one mass scale m_χ

$$\sigma \sim \frac{g^4}{m_\chi^2},$$

with g some coupling. Substitute condition (ii) into condition (i), and you will immediately find that $m_\chi \gg 0.1$ eV. As long as the cross section is "right", the cold relic mass can be in principle as low as 0.1 eV.[g] Therefore,

---

[g]Note that kinetic decoupling (see later) imposes constraints on the smallest possible mass for a thermal relic; if the relic is too light, it decouples too late, and the ensuing free-streaming erases structure to an unacceptable degree; this limits the lightest cold relic to masses above a few keV, although this is model dependent;

thermal freeze-out giving the "right" relic abundance is not peculiar to the electroweak scale, as reiterated recently in the literature (see e.g. the "WIMPless" miracle of Ref. [50]).

2.4 How heavy and how light can cold relics be?

Is there any *upper* limit to the particle dark matter mass in the cold thermal relic scheme? Indeed there is! The effective coupling constants that enter the pair-annihilation process cannot be arbitrarily large (a condition that can also be rephrased in terms of a unitarity limit in the partial wave expansion [51]; note that caveats to the unitarity argument do exist, and this limit can be evaded! (I suggest you read Ref. [51] and think about how to do that). Roughly, unitarity demands that

$$\sigma \lesssim \frac{4\pi}{m_\chi^2},$$

which implies

$$\frac{\Omega_\chi}{0.2} \gtrsim 10^{-8}\ \mathrm{GeV}^{-2} \cdot \frac{m_\chi^2}{4\pi}.$$

Therefore, having $\Omega_\chi \lesssim 0.2$ implies

$$\left(\frac{m_\chi}{120\ \mathrm{TeV}}\right)^2 \lesssim 1,$$

or $m_\chi \lesssim 120$ TeV.

Is there, similarly, a *lower* limit in the cold thermal relic scheme? We commented above on the general limit, for arbitrarily low cross sections, $m_\chi \gg 0.1$ eV. But, suppose now we have in mind a WIMP, i.e. a particle that interacts via electroweak interactions, for example a massive neutrino with

$$\sigma \sim G_F^2\, m_\chi^2$$

and $x_{\mathrm{f.o.}} \sim 20$. In this case

$$\Omega_\chi h^2 \sim 0.1 \frac{10^{-8}\ \mathrm{GeV}^{-2}}{G_F^2\, m_\chi^2} \sim 0.1 \left(\frac{10\ \mathrm{GeV}}{m_\chi}\right)^2.$$

---

additional limits stem from affecting the number of relativistic degrees of freedom, with effects on the CMB and on BBN [49]; the resulting constraints force the lightest cold thermal relics to masses in the few MeV range, typically.

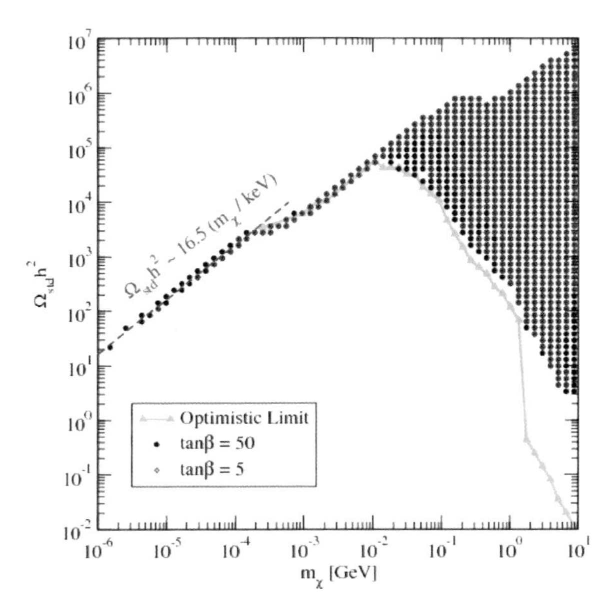

Figure 2.2: WIMP miracles? The thermal relic density of the lightest neutralino in the minimal supersymmetric extension of standard model (MSSM), as a function of the neutralino mass.
Source: Adapted from Ref. [53].

This implies that $m_\chi \gtrsim 10$ GeV for WIMPs — a limit known in the literature as the *Lee–Weinberg* limit [52]. There are a variety of reasons why such light particles cannot possibly be charged under SU(2) of weak interactions. However, Lee–Weinberg-type limits, with G_F replaced by some appropriate effective constant, apply rather broadly to a variety of WIMP models.

Consider for example the lightest neutralino in the MSSM: various effects blur a simple connection between mass and cross section/relic density, resulting in a large spread of results. The general feature of a lower limit of about 1–10 GeV is, however, rather resilient, as shown in Fig. 2.2, from Ref. [53], where I scanned generously over the MSSM parameter space. The black dots indicate points with $\tan\beta$ (the ratio of the vacuum expectation values of the two Higgses in the MSSM) fixed to 50, while the red points have $\tan\beta = 5$.[h] The x-axis indicates the lightest neutralino mass, while the y-axis is the thermal relic density calculated in a standard cosmology without entropy injection or a modified Hubble expansion rate. The green

[h] I invite the SUSY-aficionado to read my paper for details of the scan (also, see Appendix A.4 for a succinct primer on neutralino dark matter).

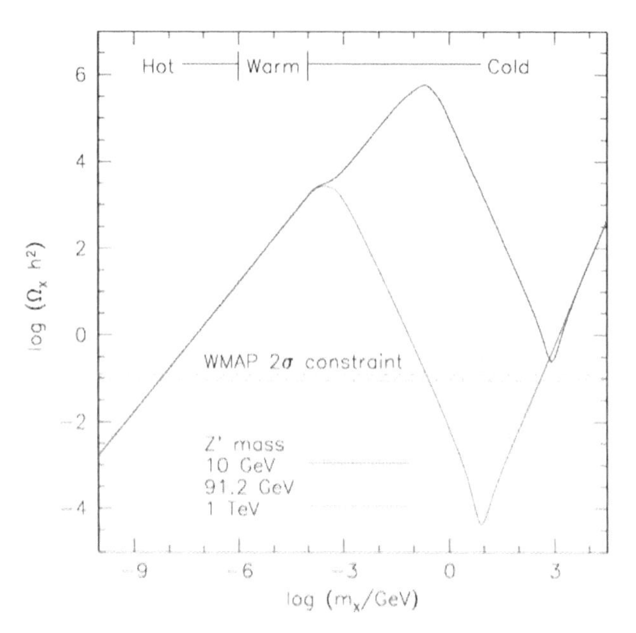

Figure 2.3: A simplified WIMP model: The thermal relic density of a relic that pair-annihilates with the cross section of Eq. (2.5), for three values of $m_{Z'}$.
Source: Adapted from Ref. [54].

line shows the envelope representing the lowest-possible neutralino thermal relic density I could find for a given neutralino mass: for $\Omega h^2 \simeq 0.1$ the lightest neutralinos I found are in the few GeV range, roughly consistent with the Lee–Weinberg limit. The figure also illustrates that the WIMP thermal relic density can vary over several orders of magnitude: the WIMP miracle is more of a *possibility* than a *firm statement*!

Figure 2.3, from Ref. [54], illustrates the thermal relic density of a WIMP as a function of the dark matter particle mass for a model with a vector-like Z' mediator of mass 10, 91.2, and 1000 GeV. The cross section is assumed to be of the form

$$\sigma \sim \frac{m_\chi^2}{(s - m_{Z'}^2)^2 + m_{Z'}^4}, \tag{2.5}$$

with s the total center-of-mass energy squared. The mass of the mediator Z' is taken to be 10 GeV, 91.2 GeV (the Z mass), and 1 TeV. The asymptotic hot and cold relic behaviors are clearly visible and match the predictions we made above: a hot relic density scales linearly with mass, a cold relic with $m_\chi \gg m_{Z'}$ has $\Omega \sim 1/\sigma \sim m_\chi^2$, and a cold relic in the regime where $m_\chi \ll m_{Z'}$

has $\Omega \sim m_{Z'}^4/m_\chi^2$. Also notice the drop at $m_\chi \sim m_{Z'}/2$, corresponding to on-shell s-channel Z' exchange, which illustrates we need to learn how to treat *resonant* pair-annihilation: we will do that in the next chapter. Finally, for $m_\chi \gg m_{Z'}$ one enters the asymptotic regime where $\sigma \sim 1/m_\chi^2$, and thus $\Omega \sim m_\chi^2$: as we will see in Sec. 9.6 this is the regime relevant for a "heavy neutrino" dark matter scenario.

Exercise 37. Explain why the three lines in Fig. 2.3 overlap exactly at $m_\chi \gg m_{Z'}$ and $m_\chi \ll m_{Z'}$.

Let us now ask the general question: how light can a WIMP be? So far, we assumed an iso-entropic universe. Suppose that after a WIMP has frozen out, and thus decoupled from the universe's thermal bath, the entropy density changes from $s \to \gamma \cdot s$, $\gamma > 1$ from e.g. decaying relics (such as relic gravitinos, moduli, ...) or from a first order phase transition (see e.g. Ref. [55]). Then

$$Y_{\rm today} \to \frac{Y_{\rm today}}{\gamma} \quad \text{and} \quad \Omega_\chi \to \frac{\Omega_\chi}{\gamma}.$$

For a sufficiently large γ, the relic abundance of almost any otherwise over-abundant relic WIMP can be "diluted" enough to match the observed dark matter density. For example, in supersymmetry the lightest neutralino can be almost arbitrarily light as long as it is bino-like (to prevent an excessively light associated chargino) and if sfermions are sufficiently heavy to suppress energy loss mechanisms in stars (for example $e^+e^- \to \chi\chi$). The additional requirement is of course that the entropy injection happens at a temperature smaller than the neutralino freeze-out, which sets a (weak) constraint on the neutralino mass [53]: in order to maintain the successful predictions of light elemental abundances, entropy injection cannot happen too close to the era of BBN, but to dilute the thermal relic it must happen after freeze-out. For a cold relic, this means

$$T_{\rm freeze\text{-}out} \sim m_{\rm cold\ relic}/(20 \cdots 50) \gg T_{\rm entropy\ injection} \gg T_{\rm BBN} \sim 1\ {\rm MeV}.$$

Any otherwise overabundant thermal relic can thus be diluted to the "right" relic density, as long as its mass is larger than 20–50 MeV.

Chapter 3

The Thermal Relic Paradigm: A Closer Look

3.1 Praeludium

The dark matter literature is flooded with the symbol $\langle \sigma v \rangle$, short for the *thermally averaged pair-annihilation cross section times velocity* — but do we really understand what this symbol indicates, and where it comes from? How is the "thermal average" defined? What is v? Is it a relative velocity? How exactly is it calculated?

Also, what if the cross section rapidly changes with energy, for example if there is a resonance? In the previous chapter we saw a simple model (a Z' mediator) where that happened. How do we deal with that, in detail? And how does the early-universe (i.e. high-temperature) cross section needed to have a "good" thermal relic compare with the late-universe (close-to-zero temperature) cross section relevant for the pair-annihilation rate today (and thus indirect dark matter detection rates)?

This chapter starts by addressing these questions in some detail, and by clarifying how the relevant calculations are performed (Sec. 3.2). In Sec. 3.3 I then discuss three famous "exceptions in the calculation of relic abundances" (as they were dubbed in the classic Ref. [56]), and also discuss, in Sec. 3.4, what happens if you fiddle around with the "right-hand side" of the freeze-out condition $\Gamma \sim H$, i.e. if you consider a modified expansion rate at freeze-out.

After the number density of a cold thermal relic freezes out, the relic continues to get "pushed around" by particles in the thermal bath. These kicks can keep the relic in *kinetic* equilibrium long after it has frozen out of *chemical* equilibrium. But, like all good things, this fun process also eventually comes to an end, and the species *kinetically decouples*. This is a defining moment in how particle properties of the dark matter affect the large-scale structure of the universe: kinetic decoupling sets the size of the *smallest halos* that can form for a given particle dark matter candidate! We will go over all of this, and compare what we find with what happens, instead, in the case of hot and warm relics, in Sec. 3.6.

Lastly, I intend to bring to the Reader's attention a few "lesser-known", but in my opinion really interesting possibilities beyond the standard thermal relic freeze-out story (Sec. 3.7). There are many good physics lessons to be learned from things like self-interacting dark matter (SIDM), elastically decoupling relics, semi-annihilation, and freeze-in.

3.2 The Boltzmann equation

The starting point for a closer look at the calculation of the relic density of species is the Boltzmann equation, that is generically cast as [13]:

$$\hat{L}[f] = \hat{C}[f], \tag{3.1}$$

where $f = f(\vec{p}, \vec{x}, t)$ is the (to-be-determined) phase-space density,[a] $\hat{L}$ is the *Liouville operator* describing the change in time of the phase-space density, and $\hat{C}$ is the *collision operator* describing the number of particles per phase-space volume lost, or gained, per unit time. For those (like me) who need a refresher on the Liouville operator, its non-relativistic form is easily understood as a "total" time derivative, and reads

$$\hat{L}_{\mathrm{NR}} = \frac{\partial}{\partial t} + \frac{d\vec{x}}{dt}\vec{\nabla}_x + \frac{d\vec{v}}{dt}\vec{\nabla}_v,$$

while its covariant form reads

$$\hat{L}_{\mathrm{cov}} = p^\alpha \frac{\partial}{\partial x^\alpha} - \Gamma^\alpha_{\beta\gamma}\, p^\beta\, p^\gamma \frac{\partial}{\partial p^\alpha}.$$

Exercise 38. The diligent Reader will show that the non-relativistic form derives from the covariant form in the non-relativistic limit.

[a]Note that here I'm using momenta instead of velocities as in Sec. 1.7.

In a homogeneous and isotropic cosmology (also known as a Friedman–Robertson–Walker universe),

$$f(\vec{x},\vec{p},t) \to f(|\vec{p}|,t) \quad \text{or, equivalently}: \quad f(E,t).$$

(Note that this is just a simple special case of Jeans's theorem, see Sec. 1.7). Additionally, $\hat{L}$ simplifies to

$$\hat{L}[f] = E\frac{\partial f}{\partial t} - \frac{\dot{a}}{a}|\vec{p}|^2\frac{\partial f}{\partial E}.$$

Here, we are interested in particle number densities, i.e. we want to integrate out momenta in the phase-space density,

$$n(t) = \sum_{\text{spin}} \int \frac{\mathrm{d}^3 p}{(2\pi)^3} f(E,t).$$

We will thus take Eq. (3.1) and consider (calling g the number of spin degrees of freedom)

$$\int L[f]\cdot g\frac{\mathrm{d}^3 p}{(2\pi)^3} = \frac{\mathrm{d}n}{\mathrm{d}t} + 3H\cdot n,$$

where we introduced the Hubble rate, $H = \dot{a}/a$, and where we have integrated by parts using

$$\frac{1}{a^3}\frac{\mathrm{d}}{\mathrm{d}t}\left(a^3\cdot n\right) = \frac{\mathrm{d}n}{\mathrm{d}t} + 3H\cdot n.$$

Cleaning up the right-hand side of the Boltzmann equation is a bit messier and I recommend the classic paper by Gondolo and Gelmini, Ref. [57]. For definiteness, let us consider a process of the type

$$1 + 2 \leftrightarrow 3 + 4,$$

where we are interested in the number density of species 1, and where we assume that species 3 and 4 are in thermal equilibrium. The right-hand side of Eq. (3.1) can then be cast as:

$$g_1 \int \hat{C}[f_1]\frac{\mathrm{d}^3 p}{(2\pi)^3} = -\langle\sigma\cdot v_{\text{M}\text{\o}1}\rangle\left(n_1 n_2 - n_1^{\text{eq}} n_2^{\text{eq}}\right),$$

where $n_{1,2}$ are the number densities, while $n_{1,2}^{\rm eq}$ indicate the equilibrium number densities, and where

$$\sigma = \sum_f \sigma_{12\to f},$$

indicates the invariant, unpolarized total cross section for processes $1+2 \to$ any final state f in thermal equilibrium, and where, finally, the "Møller velocity"[b] is defined by the following, explicitly covariant form:

$$v_{\rm Mol} \equiv \frac{\sqrt{(p_1 \cdot p_2)^2 - m_1^2 m_2^2}}{E_1\, E_2}.$$

A couple of comments:

1. Note that $v_{\rm Møl}\, n_1\, n_2$ is a Lorentz-invariant quantity;
2. Note that in the rest frame of 1 (or 2; what we can think of as the "lab frame"), $v_{\rm Møl} \to v_{\rm rel} = |\vec{v}_1 - \vec{v}_2|$, where e.g. $\vec{v}_1 = \vec{p}_1/E_1$ etc.

Last ingredient: the thermal average: this is defined by the expression:

$$\langle \sigma \cdot v_{\rm Møl} \rangle = \frac{\int \sigma \cdot v_{\rm Møl}\, e^{-E_1/T} e^{-E_2/T}\, {\rm d}^3 p_1\, {\rm d}^3 p_2}{\int e^{-E_1/T} e^{-E_2/T}\, {\rm d}^3 p_1\, {\rm d}^3 p_2}. \qquad (3.2)$$

In the expression above we have made two assumptions:

(1) that the Maxwell–Boltzmann approximation holds (as an approximation to Fermi–Dirac or Bose–Einstein statistics), and
(2) that the initial chemical potentials of species 1 and 2 are negligible.

For the calculation of the thermal abundance of cold relics, the two assumptions above are usually fine (obvious exceptions exist, such as for example for the chemical potential for asymmetric dark matter models, see Sec. 9.5).

Exercise 39. Explain why the Maxwell–Boltzmann approximation makes sense here, i.e. why at the relevant energies and temperatures for the processes we are interested in $e^{E/T} \gg 1$.

[b]Usually the Møller velocity is defined as $v_{\rm Møl} \equiv \left((\vec{v}_1 - \vec{v}_2)^2 - (\vec{v}_1 \times \vec{v}_2)^2\right)^{\frac{1}{2}}$.

Exercise 40. Evaluate the denominator of Eq. (3.2) for $m_1 = m_2$.

The diligent Reader who carried out the exercise above found that the denominator of Eq. (3.2), for $m_1 = m_2 = m$ (the pair-annihilation case relevant for us) reads

$$\int e^{-E_1/T} e^{-E_2/T} \, \mathrm{d}^3 p_1 \, \mathrm{d}^3 p_2 = \left(4\pi m^2 T K_2\left(\frac{m}{T}\right)\right)^2,$$

where K_2 is the modified Bessel function of the second order. The numerator reads, instead,

$$\int \sigma \cdot v_{\text{M}\varnothing\text{l}} e^{-E_1/T} e^{-E_2/T} \, \mathrm{d}^3 p_1 \, \mathrm{d}^3 p_2$$
$$= 2\pi^2 T \int_{4m^2}^{\infty} \sigma(s)(s - 4m^2)\sqrt{s} K_1\left(\frac{\sqrt{s}}{T}\right) \mathrm{d}s, \tag{3.3}$$

a "convolution" of the cross section σ times a factor $s - 4m^2$, where s is the center-of-mass total energy, and a temperature-dependent thermal kernel. This form is key to understand important caveats to, e.g. the "miraculous" relation that implies that $\langle \sigma v \rangle = 3 \times 10^{-26}$ cm^3/s gives the correct thermal relic density. From now on, I will suppress the subscript Møl and intend always that $v \to v_{\text{Møl}}$.

The integrals in Eqs. (3.2) and (3.3) can be cast in a rather useful and more general form by defining an "effective" annihilation rate per unit volume and unit time $W_{\text{eff}}(p_{\text{eff}})$, which is a function of a new integration variable, the effective momentum p_{eff} defined by

$$s = 4p_{\text{eff}}^2 + 4m^2,$$

and a temperature-dependent weight function $\kappa(T, p_{\text{eff}})$ [58]. The effective annihilation rate, including the possibility of co-annihilating species i, j (i.e. of other particle species participating in the freeze-out process; $i = 1$ indicates the stable species whose relic density we are calculating, and of mass $m_1 = m$), is defined as

$$W_{\text{eff}} \equiv \sum_{ij} \frac{p_{ij}}{p_{11}} \frac{g_i g_j}{g_1^2} W_{ij},$$

with g_i, as usual, the internal degrees of freedom of species i, and with

$$p_{ij} = \frac{\sqrt{s-(m_i+m_j)^2}\sqrt{s-(m_i-m_j)^2}}{2\sqrt{s}},$$

and with the annihilation rate per unit volume and unit time for the

$$i + j \leftrightarrow \text{anything}$$

processes, of cross section σ_{ij}, defined as

$$W_{ij} = 4p_{ij}\sqrt{s}\sigma_{ij} = 4\sigma_{ij}\sqrt{(p_i \cdot p_j)^2 - m_i^2 m_j^2} = 4E_iE_j\sigma_{ij}v_{ij},$$

where the Reader will recognize the Møller velocity for the ij process v_{ij} as defined above. The expression for the thermally averaged cross section, including co-annihilation, can then be cast as

$$\langle \sigma_{\rm eff} v\rangle = \int_0^\infty dp_{\rm eff}\frac{W_{\rm eff}(p_{\rm eff})}{4E_{\rm eff}^2}\kappa(p_{\rm eff},T), \tag{3.4}$$

where

$$E_{\rm eff}^2 = \sqrt{p_{\rm eff}^2 + m^2}.$$

Exercise 41. Give the explicit form for the function $\kappa(p_{\rm eff},T)$ by comparing Eqs. (3.4) and (3.2).

In Eq. (3.4), the term $W_{\rm eff}(p_{\rm eff})/4E_{\rm eff}^2$ is an "effective $\langle\sigma v\rangle$"; in the $p_{\rm eff} \to 0$ limit, for example, it reduces to the pair-annihilation cross section for the lightest species at $T = 0$, relevant for late-universe annihilation, and thus indirect detection. Incidentally, the `DarkSUSY` code [59], one of the state-of-the-art software for the calculation of the thermal neutralino relic abundance,[c] utilizes the integral in the form of Eq. (3.4) to carry out the calculation of $\langle\sigma v\rangle$.

Figure 3.1 illustrates a few example $W_{\rm eff}(p_{\rm eff})/4E_{\rm eff}^2$ functions for neutralino dark matter models, as well as the function $\kappa(p_{\rm eff},T)$, shown for

[c]The interested Reader should also be aware of a competing code, MicrOMEGAs [60], available at https://lapth.cnrs.fr/micromegas/, which is readily customizable for the calculation of the relic density of arbitrary particle theory models, via an interface to `CalcHEP` model files.

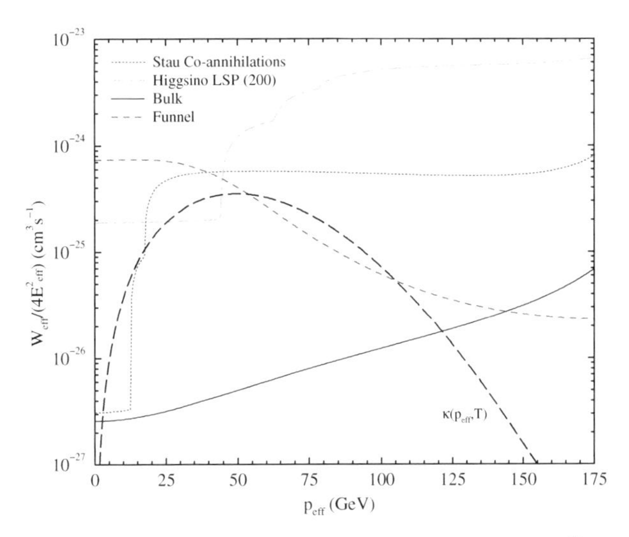

Figure 3.1: The effective annihilation cross section $W_{\rm eff}(p_{\rm eff})/4E^2_{\rm eff}$ for a few neutralino dark matter models, illustrating a resonant channel, a model with one co-annihilating partner, and a model with multiple co-annihilating partners. For details, see Ref. [61], from which the figure is adapted.

$T = T_{\rm freeze\text{-}out}$ (thus showing which $p_{\rm eff}$ values are effective sampled at freeze-out). The lower the temperature, the more $\kappa(p_{\rm eff}, T)$ weighs low values for $p_{\rm eff}$, until at $T \to 0$ it becomes proportional to a delta function centered at $p_{\rm eff} = 0$. The red, green, black, and blue curves illustrate several possibilities occurring for neutralino dark matter in the minimal supersymmetric extension of standard model (MSSM): notice a resonant annihilation channel (dashed blue line), the case of one co-annihilating partner (the scalar partner of the tau, red dotted line) and a case of multiple co-annihilating partners (dot-dashed green line). We will discuss resonances and co-annihilation in detail in the next section.

Note that in the non-relativistic limit ($x = m/T \gg 1$) it is sometimes helpful to expand $\langle\sigma v\rangle$ in inverse powers of x, using $s = 4m^2 + m^2v^2$.

Exercise 42. Suppose that

$$\langle\sigma v\rangle = \langle a + bv^2 + cv^4 + \cdots\rangle.$$

Show that

$$\langle\sigma v\rangle = a + \frac{3}{2}bx^{-1} + \frac{15}{8}cx^{-2} + \cdots$$

The $a \gg b$ limit is sometimes referred to as s-wave annihilation, while if $a \ll b$ then one has a p-wave annihilation.

Exercise 43. Calculate the coefficients a and b for a cross section $\sigma = \alpha^2/m_\chi^2$ and for a cross section $\sigma = \alpha^2 T^2/m_\chi^4$.

To sum up, we have a non-linear differential equation that looks like this:

$$\dot{n} + 3Hn = \langle \sigma v \rangle \left(n_{\rm eq}^2 - n^2 \right). \qquad (3.5)$$

The equation looks a bit better using the co-moving number density, which factors out the dilution in the physical number density due to the expansion of the universe. Let me remind you once again that since in an "iso-entropic" (no changes in entropy) universe sa^3 is conserved, where s is the entropy density and a is the scale factor (see Appendix B), then the quantity $Y = n/s$ is proportional to a co-moving number density. Also, note that, if $g_{*s}(T)$ is constant, $\dot{s} = -3sH$.

Exercise 44. Show that in fact $\dot{s} = -3sH$.

Thus, again if $g_{*s}(T)$ is constant,

$$\frac{dY}{dt} = \langle \sigma v \rangle s \left(Y_{\rm eq}^2 - Y^2 \right) \quad \rightarrow \quad \frac{dY(x)}{dx} = -\frac{xs\langle \sigma v \rangle}{H(m)} \left(Y(x)^2 - Y_{\rm eq}^2(x) \right). \qquad (3.6)$$

Exercise 45. Tease out what exactly $H(m)$ is (in this case, see Ref. [13]).

While there are no analytic solutions to Eq. (3.6) above, it is clear (see also the discussion in Appendix B.3) that defining an annihilation rate Γ as

$$\Gamma \equiv n_{\rm eq} \langle \sigma v \rangle$$

and recasting Eq. (3.6) as

$$\frac{x}{Y_{\rm eq}} \frac{dY}{dx} = -\frac{\Gamma}{H} \left[\left(\frac{Y}{Y_{\rm eq}} \right)^2 - 1 \right],$$

we have two regimes, with the boundary case $\Gamma \sim H$ defining the freeze-out temperature $T_{\rm f.o.}$: for $\Gamma \gg H$ (which, as we know, is when the particle is

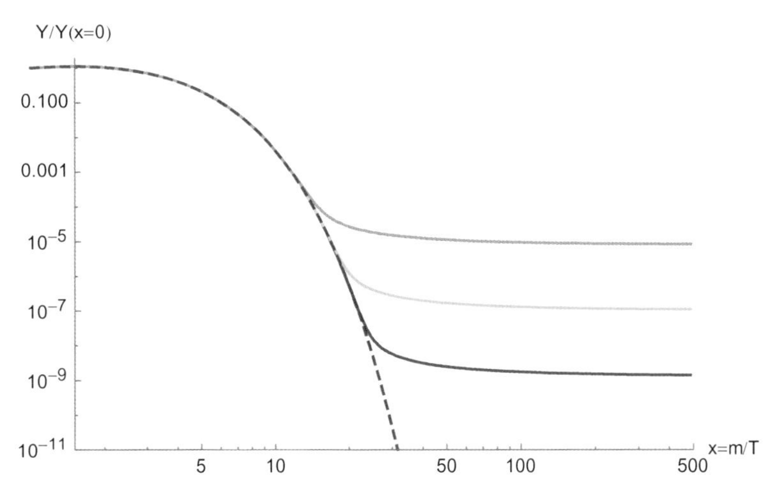

Figure 3.2: Decoupling of a thermal relic, for $\sigma = 10^{-14}$, 10^{-12}, 10^{-10} GeV^{-2}, respectively, for the orange, green, and blue curve, shown as the co-moving number density $Y = n/s$ normalized to its $T \to \infty$ (i.e. $x = 0$) value, against $x = m/T$, with $m = 1$ TeV; the red dashed curve represents the equilibrium distribution.

in equilibrium and corresponds to $x \lesssim x_{\rm f.o.}$), $Y(x \lesssim x_{\rm f.o}) \simeq Y_{eq}$, while for $\Gamma \ll H$ $Y(x \gtrsim x_{\rm f.o}) \simeq Y_{eq}(x_{\rm f.o.})$. Figure 3.2 illustrates what happens to $Y(x)$ for a few different values of σ. In particular, notice how the figure illustrates, as expected, that the asymptotic relic density scales as $1/\sigma$.

For a cold relic, after freeze-out the number density is much larger than the (exponentially suppressed) equilibrium number density. Thus we can write

$$\frac{dY}{dx} \simeq -\frac{\lambda}{x^{n+2}} Y^2, \quad \text{where} \quad \lambda = \frac{\sigma_0 s_0}{H(m)},$$

where I assumed that the cross section has the form $\langle \sigma v \rangle = \sigma_0 x^{-n}$ (of course, $n=0$ for s-wave annihilation, $n = 1$ for p-wave etc.) and $s = s_0 x^{-3}$. Then we can go ahead and solve the differential equation by separation of variables, and neglecting $1/Y_{\rm f.o.}$ versus $1/Y_{\rm today}$ we get

$$Y_{\rm today} \simeq \frac{n+1}{\lambda} x_{\rm f.o.}^{n+1}. \tag{3.7}$$

The last task is to determine (numerically) $x_{\rm f.o.}$, which is exactly what we carried out in Eq. (2.2) in the previous chapter, and which gives a logarithmic dependence on the particle mass, but whose analytic form in terms of m_χ, n, λ is ugly you will convince yourself of in the second exercise proposed below.

Exercise 46. Show that Eq. (3.7) agrees, for $n = 0$, with what we got in the last chapter in Eq. (2.3).

Exercise 47. Define the "departure from equilibrium" function

$$\Delta(x) \equiv Y(x) - Y_{\rm eq}(x),$$

and use the freeze-out condition (following [13])

$$\Delta(x_{\rm f.o.}) = cY_{\rm eq}(x_{\rm f.o.}),$$

with c a numerical constant of order unity.

(a) Show that the freeze-out criterion approximately gives

$$x_{\rm f.o.} \simeq \ln[(2+c)\lambda ac] - \left(n + \frac{1}{2}\right)\ln\{\ln[(2+c)\lambda ac]\},$$

where $a \simeq 0.145(g/g_{*s})$ (for the definition of g_{*s} see Appendix B).

(b) Following Ref. [13], and choosing $c(c+2) = n+1$, which gives a better-than 5% fit to the numerical results for the final relic density, show that

$$x_{\rm f.o.} \simeq \ln[0.038(n+1)(g/g_*^{1/2})M_P m\sigma_0]$$

$$-\left(n + \frac{1}{2}\right)\ln\{\ln[0.038(n+1)(g/g_*^{1/2})M_P m\sigma_0]\}, \text{ and}$$

$$Y_{\rm today} \simeq \frac{3.79(n+1)x_{\rm f.o.}^{n+1}}{(g_{*s}/g_*^{1/2})M_P m\sigma_0}.$$

3.3 Three classic (and important) caveats to the standard story

A classic paper by Griest and Seckel, Ref. [56], aptly titled "Three exceptions in the calculation of relic abundances" discusses three important departures from a vanilla estimate of the thermal relic density of cold relics. The three memorable exceptions are:

1. Resonances;

2. Thresholds (including "forbidden" dark matter [62]);
3. Co-annihilations.

Interestingly, these three exceptions are not real "exceptions": they appear rather ubiquitously in several highly respected models of particle dark matter. Take for example supersymmetric neutralinos (for a quick primer see Appendix A.4) — you have got all three, in copious numbers, in several cases of interest. Take universal extra dimensions (UED) (see Appendix A.5): once again, almost by construction, you have exceptions 1 and 3, but you can easily have 2 as well. Worth it then: Let us have a closer look!

1. Resonances

Resonant annihilation through a particle with the right quantum numbers and a mass $m_A \simeq 2m_\chi$, is found for example in the so-called "funnel" region of the (soon to be gone, thanks Large Hadron Collider (LHC)!) minimal supergravity/constrained MSSM model. There, the lightest neutralino is the dark matter candidate, of mass m_χ. In the "funnel" (from the shape of the parameter space allowed region) region, the mass of the heavy Higgses is close to twice the neutralino mass. Consider for example the mass of the CP-odd Higgs, m_A. In the funnel $m_\chi \simeq m_A/2$. There are two possibilities: either $m_\chi \gtrsim m_A/2$, or $m_\chi \lesssim m_A/2$.

Exercise 48. Sketch the effective annihilation cross section $W_{\rm eff}/4E^2_{\rm eff}$ for the two cases.

In the first case, the cross section peaks at $s = m_A^2$ and is thus most relevant at temperatures

$$T_{\rm res} \simeq m_A^2/(6m_\chi).$$

Exercise 49. Show that $\langle s\rangle \simeq 4m_\chi^2 + 6m_\chi T$.

If $T_{\rm res} \simeq T_{\rm f.o.}$ the resonance is extremely important at freeze-out, and hence for the thermal relic density; the pair-annihilation cross section today, the one that we care about for indirect dark matter detection rates, will however be potentially much lower! In this case, the freeze-out cross section might be close to $\langle\sigma v\rangle = 3\times 10^{-26}$ cm^3/s, but (potentially) the $T=0$ cross section $\langle\sigma v\rangle_0 \ll 3\times 10^{-26}$ cm^3/s.

If $m_\chi \lesssim m_A/2$, in the integral of Eq. (3.3) the cross section is always maximal for $T=0$, the resonance can be (or not) sub-dominant at freeze-out, but we are in the (lucky if one wants a signal, unlucky if one wants to hide it!) circumstance where the $T=0$ cross section $\langle\sigma v\rangle_0 \gg 3\times 10^{-26}$ cm^3/s. This case is illustrated by the blue dashed line in Fig. 3.1: the "weight" function κ, shown in the figure in its $T=T_{\rm f.o.}$ incarnation, samples the effective annihilation rate at values where it is *lower* than its $T=0$ (corresponding, in the figure, to $p_{\rm eff}=0$) value. The opposite case, as the Reader found in the exercise above, would have the "hump" in the $W_{\rm eff}/4E^2_{\rm eff}$ function at larger values of $p_{\rm eff}$, resulting in the lower value for the pair-annihilation today versus its effective value at $T=T_{\rm f.o.}$.

It must be noted that this discussion has model-dependent caveats: for example, in supersymmetry the lightest neutralinos are Majorana particles, and in a purely s-wave annihilation (at $T=0$) a pair of neutralinos are in a CP-odd state. Therefore, for example, the pair-annihilation via a CP-even particle (such as a neutral CP-even Higgs) cannot contribute to the $T=0$ pair-annihilation cross section!

2. Thresholds

Thresholds affect the relation between the freeze-out and $T=0$ pair-annihilation cross sections in an obvious way: the cross section in Eq. (3.3) suddenly increases as, e.g. $s > 4m_t^2$, where m_t is the particle in which pair our annihilating particle can go, $\chi\chi \to \bar{t}t$ (think e.g. of $m_\chi \lesssim m_t$ where t is the Standard Model (SM) top quark). Thresholds therefore always imply $\langle\sigma v\rangle_{\rm f.o.} > \langle\sigma v\rangle_0$.

An interesting case is when the dark matter dominantly pair-annihilates into a slightly heavier dark-sector state ϕ which is also in thermal equilibrium with the relativistic bath in the early universe. The dominant reaction that keeps the dark matter in thermal equilibrium is thus $\chi\chi \leftrightarrow \phi\phi$. The Boltzmann equation, with obvious meaning of the symbols, is

$$\dot{n}_\chi + 3Hn_\chi = -\langle\sigma v\rangle_{\chi\chi} n_\chi^2 + \langle\sigma v\rangle_{\phi\phi}(n_\phi^{\rm eq})^2.$$

The catch now is that if the ϕ is slightly heavier than the χ its number density is suppressed by

$$\frac{n_\phi^{\rm eq}}{n_\chi^{\rm eq}} = e^{-(m_\phi - m_\chi)/T}.$$

This in practice means that the $\chi\chi \to \phi\phi$ suffers from an exponential "phase-space" suppression from the fact that the ϕ is, kinematically, "forbidden" (at small enough T, of course, hence the name *forbidden dark matter* [62])! Detailed balance for the equilibrium number densities then sets

$$\langle\sigma v\rangle_{\chi\chi} = \langle\sigma v\rangle_{\phi\phi}\left(\frac{n_\phi^{\rm eq}}{n_\chi^{\rm eq}}\right)^2 \sim \langle\sigma v\rangle_{\phi\phi} e^{-2(m_\phi - m_\chi)/T}.$$

The suppression factor carries over to the final relic density, which will be *enhanced* by a factor of order

$$\frac{\Omega_\chi^{\rm forbidden}}{\Omega_\chi} \sim e^{+2(m_\phi - m_\chi)/T_{\rm f.o.}} \sim e^{2x_{\rm f.o.}\Delta},$$

where

$$\Delta \equiv \frac{m_\phi - m_\chi}{m_\chi}.$$

Clearly, this factor is a big deal even for moderately small relative mass splittings!

3. Co-annihilation

Co-annihilation occurs for particles whose freeze-out process is tangled with that of other particle species with a close-enough mass and a large-enough co-annihilation cross section so that the two freeze-out episodes are inter-connected. A key necessary condition is that this second co-annihilating species 2 have a mass such that at freeze-out the Boltzmann suppression of its equilibrium number density is not dramatic. In formulae, to beat Boltzmann we need $m_2 - m_1 \lesssim T_{\rm f.o.}$. In this case the relevant cross section is the "effective" cross section, including the appropriately Boltzmann-weighed contribution from (N) co-annihilating particles, given in Eq. (3.4). Schematically, we can think of the effect of co-annihilation as a change in the thermally averaged annihilation cross section of the form (if this is not obvious see the extensive and clear discussion of Ref. [63])

$$\langle \sigma v \rangle \rightarrow \langle \sigma_{\rm eff} v \rangle = \frac{\sum_{i<j=1}^{N} \sigma_{ij} \exp\left(-\frac{\Delta m_i + \Delta m_j}{T}\right)}{\sum_{i=1}^{N} g_i \exp\left(-\frac{\Delta m_i}{T}\right)}. \tag{3.8}$$

In the equation above, Δm_i indicates the difference in mass between particle i and the lightest particle (to which i, eventually, decays). Note that the denominator counts the effect of the additional degrees of freedom, suitably weighed.

Exercise 50. Show that the expression in Eq. (3.8) is consistent with Eq. (3.4).

Co-annihilation comes in two varieties, that I like to call "*parasitic*" and "*symbiotic*". If the additional degrees of freedom annihilate "less efficiently" than the particle whose number density we are interested in, then the co-annihilating particles will have a parasitic effect, and produce a smaller effective pair-annihilation cross section (because of the large denominator). This is the typical case with, for example, the lightest Kaluza–Klein excitation of UED (for a review see my own review! Ref. [64]; for a short overview, see my short overview! Appendix A.5). UED has a very compressed spectrum of particles (Kaluza–Klein modes of SM particles, whose mass differs from the compactification scale by loop corrections or corrections of the order of the SM particle mass) above the stable dark matter particle candidate. This large collection of particles brings many additional effective degrees of freedom, which outweigh the corresponding additional contribution to the pair-annihilation cross section, rendering UED a prototypical example of *parasitic* co-annihilation.

The opposite case, that of "*symbiotic*" co-annihilation, requires a "nimble" particle that

(i) annihilates efficiently, and that
(ii) does not carry a large number of degrees of freedom.

An example is co-annihilation of the lightest neutralino of the MSSM with the scalar partner of the tau lepton, or "stau", illustrated in Fig. 3.1 by the red dotted line: the effective annihilation rate has two significant jumps, the first at a value of $p_{\rm eff}$ corresponding to one stau becoming kinematically available, and a second bump where the two-stau processes are allowed. (Unfortunately, collider searches indicate that the days of the "stau co-annihilation region" may be counted [65]).

Exercise 51. From the definition of $p_{\rm eff}$ and from where the jumps in $p_{\rm eff}$ occur for the red dotted line in Fig. 3.1, reconstruct the neutralino-stau mass splitting for the model shown in the figure (see Table 1 in Ref. [61] for the solution!).

Other equally good examples are co-annihilation with charginos, which features multiple co-annihilating partners, as illustrated by the green dot-dashed line in Fig. 3.1, or stops, or other sfermions.

3.4 Modified thermal history: a quintessential example

So far we have dealt with exceptions to the left-hand side of the equation

$$\Gamma = n \cdot \sigma \sim H,$$

which is a succinct version of the Boltzmann equation. What happens if we fiddle around with the right-hand side instead, i.e. with H, the expansion history of the universe? To avoid self-promoting my own papers again, I will point you to the following example: a cosmology with a "quintessence" field that provides a dynamical dark energy term, whose impact on the relic density of weakly interacting massive particles (WIMPs) was first studied by Salati in Ref. [66].[d]

Let ϕ be the quintessence field, a spatially homogeneous real, scalar field. The field energy density and pressure are

$$\rho_\phi = \frac{1}{2}\left(\frac{\mathrm{d}\phi}{\mathrm{d}t}\right)^2 + V(\phi), \tag{3.9}$$

$$P_\phi = \frac{1}{2}\left(\frac{\mathrm{d}\phi}{\mathrm{d}t}\right)^2 - V(\phi). \tag{3.10}$$

Exercise 52. Prove Eqs. (3.9) and (3.10).

[d]See also Ref. [62] could not resist.

An example of a suitable potential that exhibits the desired "tracking" behavior (for appropriate initial conditions),[e] i.e. whose energy density tracks dynamically the dominant energy density component, is

$$V(\phi) = M_P^4 \exp\left(-\frac{\lambda\phi}{M_P}\right).$$

The field's equation of state $w = P_\phi/\rho_\phi$ moves from $w = +1$ in the "kination" phase, where the kinetic energy term dominates, to $w = -1$ in the "cosmological constant" phase, where V dominates. Tracking helps explain the coincidence problem $\Omega_\Lambda = \Omega_\phi \sim \Omega_M$, although fine-tuning is not eliminated (it creeps back in via the field's initial conditions).

Noting that $\rho_\phi \sim a^{-3(1+w)}$, in the kination phase $\rho \sim a^{-6}$ and therefore the universe is kination-dominated at sufficiently early times, with

$$H \sim \frac{T^2}{M_P}\frac{T}{T_{\rm KRE}} \qquad (T \gtrsim T_{\rm KRE}),$$

where $T_{\rm KRE}$ stands for the temperature of kination-radiation equality. To be relevant for the relic density of a particle species decoupling at $T = T_{\rm f.o.}$, kination must dominate before and at freeze-out, hence $T_{\rm KRE} > T_{\rm f.o.}$. However, to avoid disrupting Big Bang Nucleosynthesis (BBN), we must also require that $T_{\rm KRE} < T_{\rm BBN} \sim 1$ MeV.

In a kination-dominated universe, freeze-out works, schematically, exactly as we described in the previous section, and

$$\Omega_\chi^{\rm quint} = \frac{T_0^3}{M_P \cdot \rho_c} x_{\rm f.o.}\left(\frac{n_{\rm f.o.}}{T_{\rm f.o.}^2}\right),$$

but now the freeze-out condition reads

$$n_{\rm f.o.}\langle\sigma v\rangle \sim \frac{T^2}{M_P}\frac{T}{T_{\rm KRE}}.$$

We therefore have that

$$\frac{n_{\rm f.o.}}{T_{\rm f.o.}^2} \sim \frac{1}{M_P\,\langle\sigma v\rangle}\frac{T_{\rm f.o.}}{T_{\rm KRE}}.$$

[e]Which, incidentally, is where the fine-tuning is hidden.

To first order, the enhancement factor of the thermal relic density in the presence of quintessence over the standard thermal relic density is thus

$$\frac{\Omega_\chi^{\rm quint}}{\Omega_\chi^{\rm standard}} \sim \frac{T_{\rm f.o.}}{T_{\rm KRE}} \lesssim \frac{m_\chi}{20} \frac{1}{T_{\rm BBN}} \sim 10^4 \frac{m_\chi}{100\,{\rm GeV}}.$$

With more accurate calculations the enhancement factor is found to be potentially as large as 10^6 [62, 66]. This is a big deal, especially for indirect dark matter detection: dark matter particles with pair-annihilation rates today (and therefore indirect detection rates!) up to 10^6 times larger than those naively expected from standard thermal freeze-out can possibly originate as legitimate thermal relics with a modified expansion history!

3.5 Non-thermal dark matter production and asymmetric dark matter

So far we have discussed dark matter production under the hypotheses that (i) the dark matter is produced through processes which are originally in thermal equilibrium, and that (ii) the dark matter has zero chemical potential. However, there are many good reasons and theoretical frameworks where (i) the dark matter is produced *non-thermally*, by some out-of-equilibrium process, and where (ii) the dark matter is not its own antiparticle, and the relic dark matter density is the result of an particle–antiparticle asymmetry, much as in the baryonic sector.

We will discuss such possibilities at length in what follows, but let us quickly appreciate the gist of the simplest possible non-thermal dark matter production scenario. Suppose a particle species ψ, with $m_\psi > m_\chi$ is produced in the early universe with an abundance Ω_ψ, and that ψ decays to χ, which is the stable dark matter particle, at a temperature when χ would already be out of equilibrium. The relic density that χ inherits (up to contributions from the decay of other particle species and from thermal production etc.) is then simply

$$\Omega_\chi \simeq \Omega_\psi \frac{m_\chi}{m_\psi},$$

where the $\simeq$ sign indicates that additional effects (such as some entropy production in the decay process) can enter.

As you calculated in the previous chapter, the thermal abundance of protons and antiprotons is almost ten orders of magnitude smaller than the observed abundance. If an asymmetry is present, then the observed proton density can be inherited entirely from the asymmetry itself. The same could hold for the dark matter sector. Many have in fact entertained the possibility of "*asymmetric dark matter*" and embarked into explaining the coincidence $\Omega_{\rm DM} \simeq 5\Omega_{\rm B}$ (where DM is dark matter and B is baryonic matter) by postulating that, perhaps, $n_{\rm DM} \sim n_{\rm B}$ and that $m_{\rm DM} \sim 5$ GeV, with some mechanism that couples an asymmetry in one sector to the other sector, or variants on this theme. We discuss this possibility in some detail in Sec. 9.5.

3.6 Kinetic decoupling and the smallest dark matter halos

The dark matter thermal history in the very early universe is important not only for the calculation of the particle's relic density, but potentially also for the formation of matter structure in the universe, especially for (cold) WIMPs. In the early universe, elastic scattering processes such as $\chi f \leftrightarrow \chi f$, where f is a SM fermion, keep the dark matter particle in kinetic equilibrium even *after* chemical decoupling (i.e. when $\Gamma_{\chi\chi\leftrightarrow X} \ll H$). The reason is that the target densities for the processes that keep the dark matter in chemical versus kinetic equilibrium are vastly different after chemical decoupling:

$$\begin{aligned} \chi\chi \leftrightarrow ff \quad &\rightarrow \quad \Gamma = n_{\text{non-rel}} \cdot \sigma, \\ \chi f \leftrightarrow \chi f \quad &\rightarrow \quad \Gamma = n_{\text{rel}} \cdot \sigma, \end{aligned}$$

with $n_{\text{non-rel}}$ exponentially suppressed! Let us now estimate the kinetic decoupling temperature of a WIMP. We shall assume a "default" effective electroweak cross section

$$\sigma_{\chi f \leftrightarrow \chi f} \sim G_F^2 T^2.$$

We have to account for the fact that the WIMP is non-relativistic after chemical decoupling and that momentum transfer between the WIMP and the thermal bath becomes "inefficient", in a sense to be made quantitative with a simple estimate. The typical momentum transfer per collision is $\delta p \sim T$, while the WIMP momentum in the non-relativistic regime satisfies the relation

$$\frac{p^2}{2m_\chi} \sim T,$$

so that $p \sim \sqrt{m_\chi T}$. Momentum transfer is a stochastic process,[f] so it takes

$$N = \left(\frac{p}{\delta p}\right)^2 \sim \frac{m_\chi T}{T^2} = \frac{m_\chi}{T} \gg x_{\rm f.o.} \gtrsim 20,$$

collisions to establish kinetic equilibrium (m/T is much larger than $x_{\rm f.o.} \gtrsim 20$ because kinetic decoupling, typically, occurs *after* chemical decoupling; for an exception, see [67] and Sec. 3.7).

To calculate kinetic decoupling, we thus need to compare

$$n_{\rm rel} \cdot \sigma_{\chi f \leftrightarrow \chi f} \left(\frac{\delta p}{p}\right)^2 \sim T^3 \cdot G_F^2 T^2 \cdot \frac{T}{m_\chi} \sim H \sim \frac{T^2}{M_P}.$$

We thus find

$$T_{\rm kd} \sim \left(\frac{m_\chi}{M_P \cdot G_F^2}\right)^{1/4} \sim 30\ {\rm MeV} \left(\frac{m_\chi}{100\ {\rm GeV}}\right)^{1/4}.$$

What does this imply for the formation of structure in the early universe? As long as the dark matter is kinetically coupled to the cosmic heat bath via scattering off of SM particles, primordial density perturbations are damped by bulk and shear viscosity in the dark matter fluid. This damping term has the form [68, 69]

$$D_d(k) = \exp\left[-\left(\frac{k}{k_d}\right)^2\right],$$

where k is the co-moving scale (wavenumber) of the density perturbation, and the characteristic damping scale k_d is

$$k_d \simeq 1.8 \left(\frac{m_\chi}{T_{\rm kd}}\right)^{1/2} \frac{a_{\rm kd}}{a_0} H_{\rm kd} \simeq \frac{3.76 \times 10^7}{1\ {\rm Mpc}} \left(\frac{m_\chi}{100\ {\rm GeV}}\right) \left(\frac{T_{\rm kd}}{30\ {\rm MeV}}\right)^{1/2}, \quad (3.11)$$

where a is, as usual, the expansion factor, and with obvious meaning of the subscripts.

Exercise 53. Calculate the coefficient in the right-hand side of the last equality in Eq. (3.11).

After kinetic decoupling, damping no longer occurs due to the viscosity of the dark matter fluid, which is now decoupled from the cosmic heat bath.

---

[f] It undergoes, that is, Brownian motion in momentum space.

However, dark matter particles in this epoch continue to *free stream* from regions of high density to regions of low density, smoothing out inhomogeneities. The characteristic scale associated with this damping process approaches a constant value after matter-radiation equality, $T < T_{\rm eq}$, when the characteristic co-moving scale is [69].

$$k_{\rm fs} \simeq \left(\frac{m_\chi}{T_{\rm kd}}\right)^{1/2} \frac{a_{\rm eq}/a_{\rm kd}}{\ln(4a_{\rm eq}/a_{\rm kd})} \frac{a_{\rm eq}}{q_a} H_{\rm eq}.$$

Exercise 54. Show that $k_{\rm fs} \ll k_{\rm d}$.

The exercise you just carried out should have convinced yourself that density fluctuations are suppressed primarily because of free-streaming. The smallest primordial collapsed halo mass allowed by free-streaming corresponds to the dark matter mass contained in a sphere of radius $\pi/k_{\rm fs}$,

$$M_{\rm fs} \simeq \frac{4\pi}{3}\rho_\chi(T_{\rm kd})\left(\frac{\pi}{k_{\rm fs}}\right)^3.$$

Exercise 55. Give an expression for $M_{\rm fs}$ in terms of $T_{\rm kd}$ and m_χ. For the solution, see Eq. (12) in Ref. [70].

However, this is not the whole story. Refs. [71] and [72] pointed out that there is one additional source of damping: acoustic oscillations of the thermal bath itself. Such oscillations, remnants of the inflationary epoch, couple to the modes of oscillation in the dark matter fluid with co-moving wavenumbers k large enough that they enter the horizon before kinetic decoupling. These dark matter fluid modes then oscillate with the acoustic (thermal bath) modes and are damped out, while modes with k values corresponding to distance scales *larger* than the horizon size at kinetic decoupling do not experience such damping (and grow logarithmically). The damping scale for acoustic oscillations is just the horizon size at kinetic decoupling,

$$\frac{\pi}{k_{\rm ao}} \simeq \frac{1}{H_{\rm kd}}.$$

The mass scale corresponding to perturbations of wavenumber $k_{\rm ao}$ will then correspond to a cut-off scale in the matter power spectrum. This cutoff scale is the dark matter mass within the size of the horizon at kinetic decoupling, so

$$M_{\rm ao} \sim \frac{4\pi}{3}\left(\frac{1}{H(T_{\rm kd})}\right)^3 \rho_{\rm DM}(T_{\rm kd}) \sim 30\, M_\oplus \left(\frac{10\ {\rm MeV}}{T_{\rm kd}}\right)^3,$$

where $M_\oplus \simeq 3 \times 10^{-6} M_\odot$ is the mass of the Earth.

Exercise 56. Convince yourself that indeed $M_{\rm cutoff} \sim 30\, M_\oplus$ is the cutoff scale for a 10 MeV kinetic decoupling temperature.

Depending on the particle physics model, either $M_{\rm ao}$ or $M_{\rm fs}$ can be larger, and the cutoff scale to the mass of collapsed dark matter structure is set by the larger one of the two,

$$M_{\rm cutoff} = \max[M_{\rm fs}, M_{\rm ao}].$$

For typical WIMPs, these "protohalos" (which correspond to the first structures that gravitationally collapse in the early universe) have a mass comparable to the Earth mass (i.e. roughly $10^{-6}\, M_\odot$). However, in specific theories the range of variations can be very significant [73]. This fact has potentially important consequences for indirect detection, as it feeds into the problem of calculating the *boost factor* (i.e. the additional contribution to the count of pairs in a halo on top of the smooth halo component) from substructure, and for the dark matter "small-scale" problem. Specifically, the cutoff scale above should be compared with upper limits from Lyman-α data, which are in the vicinity of $M_{\rm cutoff} \ll M_{{\rm Ly}-\alpha} \simeq 10^{10}\, M_\odot$.

Exercise 57. Calculate the kinetic decoupling temperature such that $M_{\rm cutoff} \simeq M_{{\rm Ly}-\alpha}$.

As you just found, for $T_{\rm kd} \sim 0.1$ keV the cutoff to the halo mass function corresponds to the largest possible value compatible with Lyman-α data. Anything in the vicinity of a keV will affect small-scale structure in a potentially detectable way.

This is a good point to pause and estimate the cutoff scale for hot (or warm) dark matter candidates. These, remember, are defined as thermal relics ν that decoupled when relativistic, i.e. when $T \gg m_\nu$. Gravitational clustering can start when the temperature drops to $T \sim m_\nu$. The distance the hot (or warm) relic has traveled ("free-streamed") up to that point is of the order of the horizon size at that temperature, or $d_\nu \sim H^{-1}(T \sim m_\nu)$, which we can estimate as

$$d_\nu \sim \frac{M_P}{m_\nu^2}.$$

The cutoff scale is then, just as above, the dark matter mass within one horizon size (we will still use $n_\nu \sim T_\nu^3$):

$$M_{\rm cutoff,\ hot} \sim \left(\frac{1}{H(T=m_\nu)}\right)^3 \rho_\nu(T=m_\nu) \sim \left(\frac{M_P}{m_\nu^2}\right)^3 m_\nu \cdot m_\nu^3 = \frac{M_P^3}{m_\nu^2}.$$

Let us now plug in some numbers for a hot relic ($m_\nu \sim 30$ eV) and for a warm relic ($m_\nu \sim 1$ keV):

$$\frac{M_P^3}{m_\nu^2} \sim 10^{15}\, M_\odot \left(\frac{m_\nu}{30\ \text{eV}}\right)^{-2} \sim 10^{12}\, M_\odot \left(\frac{m_\nu}{1\ \text{keV}}\right)^{-2}.$$

The simple estimate indicates that a hot relic with $\mathcal{O}(10)$ eV mass cuts off structures up to cluster-size objects, which is badly at odds with observation. keV-mass "warm" dark matter, instead, produces cut-off scales on the same order of the size of Galactic halos.

Exercise 58. Show that using the naive estimate described here, Lyman-α data would imply $m_\nu \gtrsim 10$ keV.

Is there any way to probe the size of the dark matter small-scale cutoff? In Ref. [70] we pointed out that if $f = q$ (i.e. the particles the dark matter scatters off of are quarks), the processes relevant for kinetic decoupling are exactly the same as those participating in dark matter direct detection.Therefore, in principle, one could correlate $M_{\rm cutoff}$ with $\sigma_{\rm direct\ det}$. There is, however, an important caveat: usually, $T_{\rm kd} \ll \Lambda_{\rm QCD}$, the latter symbol indicating the temperature corresponding to the quantum chromodynamics (QCD) confinement phase transition. After confinement, the number density of hadrons is negligible compared to light leptons, and the latter dominate and control kinetic decoupling.[g] If, however, one has a

[g]Scattering off of hadrons could be relevant for $T_{\rm kd} \sim m_\pi$, and can be estimated e.g. via chiral perturbation theory [70].

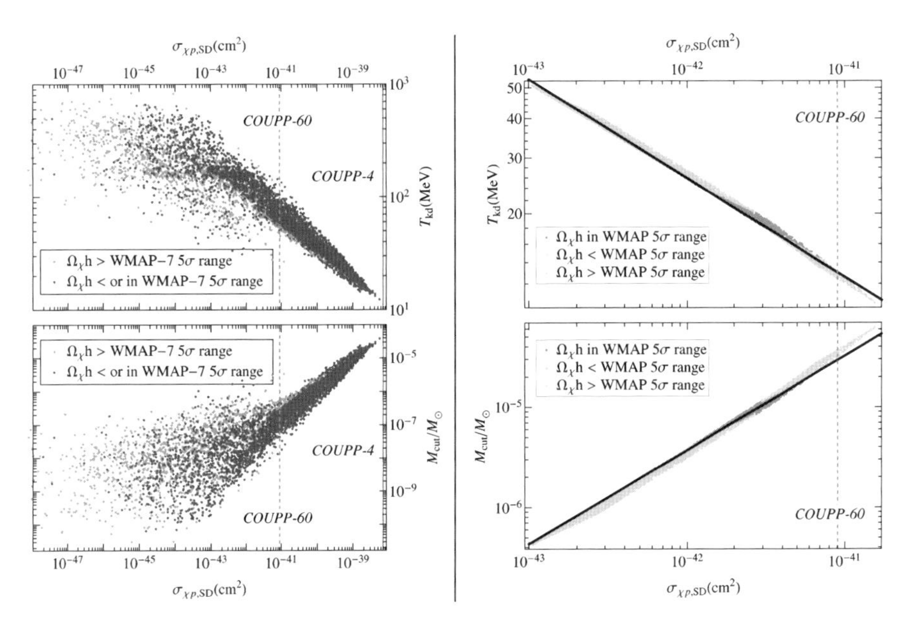

Figure 3.3: The correlation between the spin-dependent dark matter-proton cross section and the kinetic decoupling temperature (upper panels) and the small scale cutoff mass (lower panels) in the MSSM (left panels) and in UED (right panel). From Ref. [70].

dark matter theory where "quark-lepton universality" holds, then a correlation is expected, and indeed is found — both for UED and for supersymmetric dark matter [70] as illustrated in Fig. 3.3.

3.7 Lesser known stories and miracles of SIDMs, SIMPs, ELDERs, semi-annihilation, freeze-in, and FIMPs

While the standard chemical decoupling of cold relics dominates the lore in the literature of dark matter production in the early universe, by no means is this the only possibility. There are interesting lessons to be learned, and practice to be made, out of other possibilities.

SIDM

In a classic paper, Ref. [74], it was argued that dark matter candidates can be classified in terms of whether or not their co-moving number N and/or entropy S changes as the temperature drops below the (dark matter) particle mass. Hot dark matter candidates are an example where neither N nor S changes when $T \sim m_\chi$. On the other hand, for cold relics both N and S are changing.

> **Exercise 59.** Explain the statements above about N and S for hot and cold relics.

What if there were processes such that N changed, but S did not? An example would be a scenario where self-annihilation processes of the following type exist:

$$\chi\chi \to \underbrace{\chi \cdots \chi}_{n}, \quad n \geq 3. \tag{3.12}$$

If the dark matter sector decouples from the thermal bath *before* (i.e. at a higher temperature than when) processes in Eq. (3.12) go out of equilibrium, those processes will change the number density N, while no heat is exchanged with the thermal plasma, and thus S is unaffected! There are simple realizations of this self-interacting dark matter (SIDM) scenario (see also the discussion below). These scenarios generically predict significant heating of dark matter structures with important effects on (possible issues concerning) small-scale structure [75,76]; unfortunately, at least in its original incarnation, this setup predicts very light dark matter particles (masses $\lesssim$ 100 eV) which are not consistent with large-scale structure constraints.

> **Exercise 60.** Can you think of an example of the last possibility, i.e. a scenario where at $T \sim m_\chi$ S changes but N does not?

The strongly interacting massive particle (SIMP) miracle

A simple modification to the SIDM story is to allow for a small coupling with the (SM) thermal bath, such that the two sectors are in kinetic equilibrium [77], and yet $n \to 2$ ($n \geq 3$) processes effectively produce self-interacting effects, and set the thermal relic density. This possibility is known as the SIMP "miracle" [77]. The word "miracle" indicates that there exists a combination of couplings and masses that naturally explains the observed cosmological abundance of dark matter, and that such combination might be connected with the scales of QCD. Let us estimate what this combination is.

Suppose the dominant number-changing process is $3 \to 2$ ($4 \to 2$ is also possible, but more involved to realize in practice [77]). The dark matter χ number-changing rate is, here,

$$\Gamma_{3\to 2} = n_\chi^2 \langle \sigma_{3\to 2} v^2 \rangle,$$

as opposed to the usual $\Gamma_{2\to 2} = n_\chi \langle \sigma_{3\to 2} v \rangle$. For definiteness, let us assume that the cross section (by dimensional analysis) is

$$\langle \sigma_{3\to 2} v^2 \rangle \sim \frac{\alpha_{3\to 2}^3}{m_\chi^5},$$

where $\alpha_{3\to 2}$ is some effective, dimensionless coupling. Now the calculation proceeds similarly to what we did in the standard $2 \to 2$ case: find the freeze-out number density from the freeze-out condition $\Gamma_{3\to 2}(T) = H(T) \sim T^2/M_P$, which gives

$$n_{\rm f.o.} \sim T_{\rm f.o.} \left(\frac{m_\chi^5}{M_P\, \alpha_{3\to 2}^3} \right)^{1/2},$$

and use conservation of entropy, just as in Eq. (2.3). In this case, we get

$$\Omega_\chi = \frac{m_\chi\, n_{\rm f.o.}}{\rho_c\, T_{\rm f.o.}^3} = \frac{T_0^3}{\rho_c} \frac{m_\chi^{3/2}\, x_{\rm f.o.}^2}{(M_P\, \alpha_{3\to 2}^3)^{1/2}},$$

giving a preferred mass range, for a strongly self-interacting particle, $\alpha_{3\to 2} \sim 1$,

$$\left(\frac{\Omega_\chi}{0.2} \right) \sim \left(\frac{m_\chi}{35\ \text{MeV}} \right)^{3/2} \left(\frac{x_{\rm f.o.}}{20} \right)^2 \left(\frac{1}{\alpha_{3\to 2}} \right)^{-3/2},$$

a combination that might be suggestive of the typical scales of QCD.

Exercise 61. Carry out the same calculation for $4 \to 2$ processes, assuming

$$\langle \sigma_{4\to 2} v^3 \rangle \sim \frac{\alpha_{4\to 2}^4}{m_\chi^8},$$

and show that for $\alpha_{4\to 2} \sim 1$ a good thermal relic has a mass of $m_\chi \sim 100\,\text{keV}$.

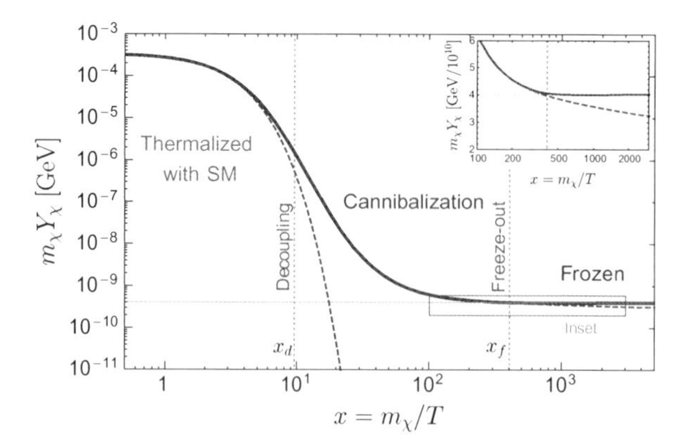

Figure 3.4: The ELDER scheme: chemical decoupling of the dark matter from the thermal bath (the dashed blue line indicates the equilibrium number density, *followed* by kinetic decoupling (indicated by x_d) and followed by self-annihilation freeze-out ($x_{\rm f.o.}$). The red dashed line shows the case where the "cannibalization" induced by self-annihilation continues after decoupling (i.e. the dark matter keeps being in chemical equilibrium with itself).
Source: Adapted from Ref. [67].

Exercise 62. Assume $\sigma_{2\to 2}$, the actual self-interaction cross section, scales as $\alpha^2_{2\to 2}/m^2_\chi$, and that $\alpha_{2\to 2} \sim 1$. Calculate the range of masses for which $\sigma_{2\to 2}/m_\chi \sim 1$ cm^2/g.

Among the interesting implications of SIMPs, besides addressing some of the small-scale structure problems such as the core versus cusp and "too big to fail" [75, 76] (see the exercise above), is that the coupling with the visible sector that keeps the dark matter in kinetic equilibrium implies some possible signals at direct and indirect searches [77].

ELastically DEcoupling Relics: ELDERs

One more possibility is that of "ELastically DEcoupling Relics", cleverly dubbed ELDERs in Ref. [67]: in this scheme, sketched in Fig. 3.4 above, after *chemical* decoupling of the dark matter from the thermal plasma, the dark matter is still in (i) *kinetic* equilibrium with the plasma, and (ii) in *chemical* equilibrium with *itself* in the presence of SIMP-like $n \to 2$, $n \geq 2$ (say, $n = 3$) processes. The key event that drives the thermal relic density is, here, kinetic decoupling, after which the dark matter enters a "cannibalization" phase driven by the fast $3 \to 2$ processes. After the latter also go out of equilibrium, the dark matter is, eventually, completely decoupled, and its co-moving number density is frozen.

Exercise 63. Assume $\sigma_{3\to 2} \sim \alpha^3_{3\to 2}/m^5_\chi$ and that the elastic scattering cross section goes wither like $\sigma_{\rm el} \sim \alpha_{\rm el}/m^2_\chi$ or like $\sigma_{\rm el} \sim \alpha_{\rm el}T^2/m^4_\chi$. Estimate, for $m_\chi \sim$10 MeV, the condition on $\alpha_{3\to 2}$ and $\alpha_{\rm el}$ for kinetic equilibrium to happen at a *higher* temperature than self-annihilation decoupling.

Semi-annihilation

Another non-standard possibility is that of *semi-annihilation* [78]: suppose the dark matter sector included several stable components χ_i. Then processes like

$$\chi_i \chi_j \to \chi_k \phi,$$

where ϕ is a SM particle or an unstable particle that decays to SM particles, would change the dark matter number density by *one unit*, contrary to the annihilation processes that changes it by two units.[h] A concrete, simple realization is a complex scalar field χ stabilized by a Z_3 symmetry: in this case both processes

$$\bar{\chi}\chi \to \phi\phi; \qquad \chi\chi \to \bar{\chi}\phi,$$

are allowed and contribute to the dark matter thermal relic density. In this case, the relevant Boltzmann equation is

$$\frac{dn_\chi}{dt} + 3Hn_\chi = -\langle\sigma v\rangle_{\bar{\chi}\chi\to\phi\phi}\left(n^2_\chi - (n^{\rm eq}_\chi)^2\right) - \frac{1}{2}\langle\sigma v\rangle_{\bar{\chi}\chi\to\bar{\chi}\phi}\left(n^2_\chi - n_\chi n^{\rm eq}_\chi\right),$$

that can be solved in the standard way taking care of including the new second term. Semi-annihilation can drive the relic density of the dark matter, and produce distinct signals in indirect detection from the ϕ decay, different from those one would have if only $\bar{\chi}\chi \to \phi\phi$ reactions were allowed [78]. For example, semi-annihilation allows for the production of photon lines from single-photon production, via the process $\bar{\chi}\chi \to \phi\gamma$. Notably, such annihilation rate is parametrically larger than $\bar{\chi}\chi \to \gamma\gamma$, since it comes with a single power of $\alpha_{\rm em}$ instead of two, as first pointed out in Ref. [79].

[h] As long as the triangle inequality $m_k < m_i + m_j$ and its crossed versions are satisfied, of course.

Freeze-in and the feebly-interacting massive particle (FIMP) miracle

Since we have been talking so much about *freeze-out*, it only seems fair to mention and describe the possibility of *freeze-in*, which is the mechanism by which very respectable dark matter candidates (e.g. sterile neutrinos, more on these guys in Chapter 8) are possibly produced in the early universe. As we will see shortly, in a well-defined sense *the yield from freeze-in is in effect the inverse of the yield from freeze-out*! In addition, there are rather interesting phenomenological consequences of freeze-in worth bearing in mind.

Suppose there is a feebly-interacting massive particle (FIMP, as dubbed in Ref. [80]) which shares a highly-suppressed interaction vertex (with some dimensionless coupling $\lambda \ll 1$) with particles in the thermal bath, and zero initial abundance. The FIMP, by construction, is assumed to never be in thermal equilibrium — its interaction with other particles being highly suppressed. Let us indicate with the acronym LOSP the "lightest observable sector particle", of mass m, in the thermal bath — the lightest particle that is in thermal equilibrium and that shares with the FIMP an unbroken symmetry (that makes the FIMP stable), and that participates in the λ vertex. Collisions or decays involving the FIMP and the LOSP produce out-of-equilibrium FIMPs, that add up over the history of the universe up to when the LOSP number density exponentially decreases at and below $T \lesssim m$. The density of "*frozen-in*" FIMPs over a Hubble time in radiation domination and for $T \gtrsim m$ is

$$Y(T) = n_{\rm FIMP}/s \sim (\Gamma_{\rm LOSP \to FIMP} t_H) \frac{n_{\rm LOSP}}{s} \sim \lambda^2 T \frac{M_P}{T^2} = \lambda^2 \frac{M_P}{T}.$$

Since for $T \lesssim m$ the LOSP number density exponentially decreases, FIMPs are dominantly produced at $T \sim m$, thus

$$Y_{\rm f.i.} \sim \lambda^2 \frac{M_P}{m}.$$

Now, let us remind ourselves what we had for freeze-out, for a particle of mass m' and pair-annihilation σ'

$$Y_{\rm f.o.} \sim \frac{1}{M_P m' \sigma'} \sim \frac{1}{(\lambda')^2} \frac{m'}{M_P},$$

where I assumed that $\sigma' \sim (\lambda')^2/(m')^2$. The yield from freeze-in and freeze-out have *inverse dependence* on the coupling and mass of the relevant particles! This is not hard to understand: for example, the bigger the coupling (and thus the bigger the relevant cross section), the more suppressed the thermal relic abundance from freeze-out, but the more efficiently particles are produced in the freeze-in scheme.

The WIMP "miracle" is the statement that if $\lambda' \sim 1$, then one has just about the right freeze-out thermal relic abundance if $m' \sim v$, where v is the scale of weak interactions. Vice versa, this implies that FIMPs are "miraculous" if, for LOSP at the weak scale, $m \sim v$, the feeble coupling is set by:

$$\lambda \sim \frac{v}{M_P} \qquad \text{FIMP miracle.}$$

There are numerous theory examples where freeze-in is realized rather naturally [80]. Additionally, there are interesting possibilities where the LOSP not only produces FIMPs via freeze-in, but also via late-time decays

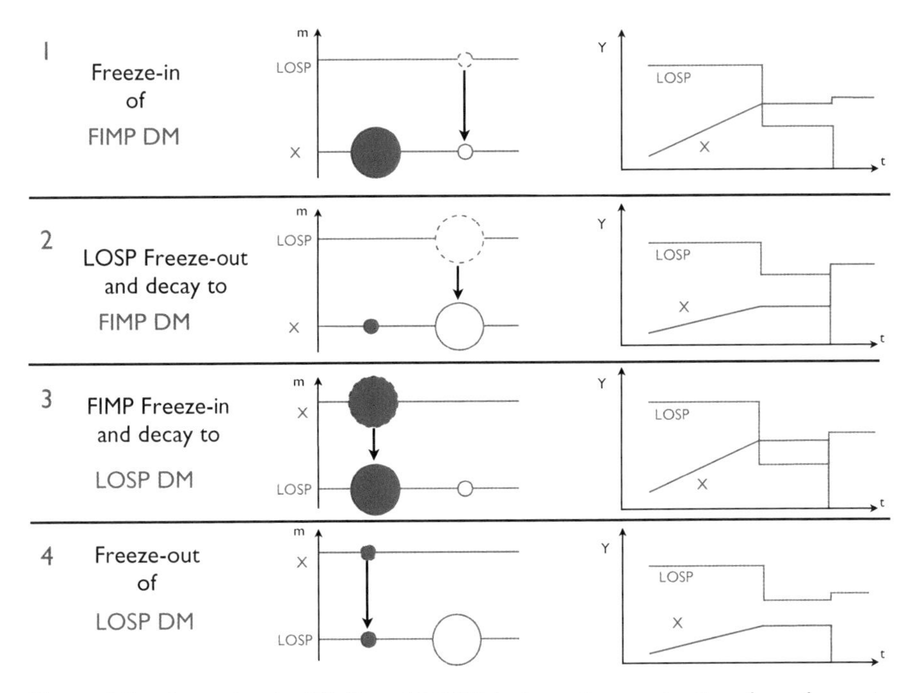

Figure 3.5: Scenarios for FIMP and LOSP dark matter production from freeze-in and freeze-out — on the left large (small) circles represent the dominant (subdominant) production mechanism for the relic dark matter abundance, while on the right is a sketch of the co-moving number densities of LOSP (blue) and FIMP (red).

Source: Adapted from Ref. [80].

(in other words, some FIMPs are produced via freeze-in at $T \gtrsim m$, where m is the LOSP mass, but then, after the LOSP freezes out, at, say $T \lesssim m/20$, the LOSP also *decays* to the FIMP, producing an additional non-thermal population out of equilibrium); The LOSP might also be lighter than the FIMP, and can thus decay into the LOSP after being produced by freeze-in. Figure 3.5 illustrates schematically the four possible cases.

Freeze-in has very interesting experimental implications. The LOSP might be produced at colliders, and if it is to produce enough FIMPs for them to be the dark matter, it is generally long-lived on collider scales and can leave interesting signals in the detector [81]; if the LOSP is the dark matter, and its relic density is set primarily by FIMP decays (case 3 in the figure), then the LOSP pair-annihilation cross section is much larger than the typical freeze-out "magic" cross section of $\langle \sigma v \rangle \sim 3 \times 10^{-26}$ cm^3/s, and annihilation rates will be very large, with very promising indirect detection prospects [80].

Chapter 4

The Art of WIMP Direct Detection

4.1 Praeludium

Directly detecting a *weakly interacting* particle is not easy. In 1934, folks like Hans Bethe and Rudolf Peierls, who had just estimated the neutrino interaction cross section to be less than "$10^{-44}\mathrm{cm}^2$ (corresponding to a penetrating power of 10^{16} km in solid matter)" [82] very optimistically quipped that: "It is therefore *absolutely impossible*[a] to observe processes of this kind with the neutrinos created in nuclear transformations" [82].

> **Exercise 64.** Which density "solid matter" has in Bethe and Peierls' estimate above?

Detecting a *new* yet-undiscovered fundamental particle that is also, at best, *weakly* interacting is, come to think about it, even harder compared to neutrinos: we do not know the hypothetical particle mass, and thus are ignorant about both the expected particle flux and the expected kinematics of the process; unlike neutrinos, we do not know the details of the particle's interaction properties, or the level at which it should interact.

Richard Crane, in 1948, stated that "Not everyone would be willing to say that he believes in the existence of the neutrino, but it is safe to say that there is hardly one of us who is not served by the neutrino hypothesis. [...] While the hypothesis has had great usefulness, it should be kept in the back of one's mind that it has not cleared up the basic mystery, and

[a]My emphasis.

that such will continue to be the case until the neutrino is somehow caught [...]" [83]. It is tempting, and fitting, almost seventy years later, to change (*mutatis mutandis*) the word "neutrino" for the word "dark matter" in the quote.

As the Reader is well aware of, neutrinos were in fact detected less than two decades after Bethe and Peierls' pessimistic statement, at the Savannah River reactor, by Reines and collaborators — a "probable detection" in 1953 [84] followed by detections at a rate of 3 per hour by 1956 [85]. The reaction that led to the discovery of neutrinos is an *inelastic* scattering process, known as the inverse beta process:

$$\bar{\nu}_e + p \to e^+ + n,$$

whose cross section is:

$$\sigma_{\bar{\nu}_e + p \to e^+ + n} \approx 10^{-43}\ (E_\nu/\mathrm{MeV})^2\ \mathrm{cm}^2. \tag{4.1}$$

The notion of *inelastic* scattering might actually apply to detecting (directly) dark matter particles. However, this comes at the price of having a non-minimal dark matter sector (similarly to the Standard Model's (SM) leptonic sector, comprising the neutrino and the corresponding charged lepton). For inelastic processes to occur, the dark sector should contain, in addition to the dark matter particle χ, a second dark particle species χ' which χ could inelastically scatter into, via processes of the type

$$\chi + X \to \chi' + Y,$$

where X and Y indicate SM particles. This possibility has been explored theoretically (see especially Ref. [86] for a neat model-independent formulation), and, for example, advocated to explain the DAMA/LIBRA conundrum [87].

Exercise 65. Show that the maximal mass splitting δ between χ' and χ to have an inelastic scattering event, as a function of the target nucleus mass m_N, of the dark matter particle mass m_χ, and of the dark matter-nucleus relative velocity $v \ll c$, is

$$\delta = \frac{v^2 m_\chi m_N}{2(m_\chi + m_N)}.$$

Calculate δ for the CDMS target nucleus Ge, for DAMA's NaI and for XENON, and check your results with Ref. [87].

If no inelastic scattering is possible, the direct detection of dark matter must proceed "elastically", i.e. via $\chi + X \to \chi + X$, or $\chi + X \to \chi + Y$. This is more difficult. Again, think about neutrinos. Elastic scattering of ν_μ off of electrons was first detected with the Gargamelle bubble chamber in 1973 [88], almost two decades after the (inleastic) direct neutrino detection. Incidentally, this was a fundamental turning point in particle physics: with that discovery we learned about weak neutral currents, thus constraining the structure of possible UV completions of the low-energy Fermi four-fermion theory (see Appendix A).

Suppose now that the dark matter particle interacts weakly (i.e. via weak interactions). A reasonable Ansatz for the form of the interaction cross section is to take Eq. (4.1) and evaluate it at some value for "E_ν". As we learned in KCM,[b] the mass scale in a two-body problem is the *reduced mass* (here, the reduced mass of the dark matter, m_χ, and nucleon m_n, proton or neutron); also, given that we think the dark matter is a non-relativistic particle today in our galaxy, its velocity is small enough that it makes sense to just use the reduced mass for the energy scale of the weak cross section; as long as $m_\chi \gtrsim m_n \sim 1$ GeV, we get

$$\sigma_{\chi n} \approx 10^{-37}\ \text{cm}^2.$$

Exercise 66. Estimate the cross section $\sigma_{\chi n}$ using G_F instead of Eq. (4.1).

Now, the scattering cross-section off of *nuclei* $\sigma_{\chi N}$ scales with that off of *nucleons* depending on the type of interaction (we will give a rationale for this below): if the dark matter couples to *nucleon number A*, the scattering is coherent on the nucleus, and, as we will see below in detail,

$$\sigma_{\chi N} \approx \sigma_{\chi n} A^2,$$

otherwise, the dark matter couples to nucleon spin, and, since that spins add incoherently inside nuclei, $\sigma_{\chi N} \approx \sigma_{\chi n}$. The former case is known as *spin-independent* direct detection, while the latter as *spin-dependent*. Below, besides explaining why these two types of coupling in the non-relativistic regime relevant here are likely the most important ones, we will attempt to spot by eye whether a theory prefers spin-dependent or independent

[b]Kindergarten Classical Mechanics :-).

scattering, a very important exercise. But first, let us think about whether or not detecting a weakly interacting particle is feasible at all.

4.2 Why direct detection is a tough business

What is the rate for dark matter scattering off of nuclei? What is the typical energy deposition we need to measure? Which dark matter particle mass can we hope to detect with a detector with a certain given energy threshold? And, really, do we have any chance to detect this scattering? Well, let us put in the numbers!

To my knowledge the first to "put in the numbers" were Goodman and Witten in 1985 [89]. Let us start with some simple (non-relativistic) kinematics: the maximal recoil momentum for a dark matter with velocity v and mass m_χ is $2m_\chi v$, and the maximal recoil kinetic energy is

$$E_{\max} = (2m_\chi v)^2/(2m_N),$$

where m_N is the recoiling target nucleus mass. This already allows us to answer the question of how light a dark matter particle we could possibly detect with a given target nucleus, assuming we have a detector with an energy threshold $E_{\rm th}$ and a maximal dark matter particle velocity $v = v_{\rm esc} \approx 500-700$ km/s.

Exercise 67. Find the minimal dark matter particle mass for a detector with He target and $E_{\rm th} = 0.1$ keV, and for a Germanium detector with a threshold $E_{\rm th} = 2$ keV, assuming $v_{\rm esc} \approx 700$ km/s.

The diligent Reader has convinced herself that it is hard to detect dark matter particles much lighter than a fraction of a GeV;[c] also, conversely, the equation above informs us that the typical energy deposition for the direct detection of weakly interacting massive particles (WIMPs) is in the keV range.

[c]Of course lowering significantly the energy threshold would change this conclusions — see for example work on scattering off of electrons in the target material, producing e.g. single-electron ionization signals [90, 91], or the interesting recent ideas put forth in Ref. [92] of using "superconducting detectors for super-light dark matter" which would have milli-eV energy thresholds!

Now, let us assume that we have a massive enough dark matter particle, and a detector with an energy threshold down to fractions of a keV; let us estimate the rate of scattering for 1 kg of detector mass and for a given target nucleus, $R = K\phi\sigma$. σ is the scattering cross section off of the given nucleus, $\phi = v\rho_{\rm DM}/m_\chi$ is the dark matter particle flux and, finally, $K \simeq 6.0 \times 10^{26}/A$ is the number of target nuclei per kilogram, each with A nucleons. The counting rate per kg per day is then

$$R = \frac{0.06 \text{ events}}{\text{kg day}} \left(\frac{100}{A}\right) \left(\frac{\sigma}{10^{-38}\text{ cm}^2}\right) \left(\frac{\rho_{\rm DM}}{0.3\text{ GeV/cm}^3}\right) \left(\frac{v}{200\text{ km/s}}\right). \tag{4.2}$$

Exercise 68. Convince yourself that the 0.06 events per kg per day is in fact right with the numbers in Eq. (4.2).

Is this something *detectable*? As always, there are two necessary conditions to have a statistically significant detection:

(i) having enough events (which one can get by building large detectors), and
(ii) having a large enough signal-to-noise (which one can get by building "smart" detectors and placing them in "smart" places).

The key question is which and how large the background (i.e. literally anything that can be confused for the signal) is.

Let us attempt to estimate how good experimenters in this field must be at beating obvious sources of background, such as radioactivity. We will use numbers from a classic 1984 paper by Druckier and Stodolsky [93] that historically set the stage for this type of estimate, in connection with neutral current detection. Consider two key types of radioactive contaminants: slowly-decaying "primeval" nuclides (e.g. U, Th, or ^{40}K) and very rare but fast-decaying trace elements such as tritium or ^{14}C.

Exercise 69. Given a typical relative ^{40}K abundance of 10^{-4}, and the ^{40}K half-life of 1.3×10^9, calculate the ^{40}K suppression factor (i.e. purity times rejection rate) required to bring the background rate to the level of 0.01 events per kg per day.

Exercise 70. Given a typical relative ^{3}H abundance of 5×10^{-18}, and the ^{3}T half-life of 12 yr, calculate the ^{3}H suppression factor (i.e. purity times rejection rate) required to bring the background rate to the level of 0.01 events per kg per day.

The two exercises above should have convinced the diligent Reader that background suppression in the business of direct detection is indeed a highly non-trivial feat, but one that has achieved incredible (at least to a theorist) levels.

Other than suppressing background, one can think of other handles that could help distinguish a *signal* event (or a collection thereof) from a *background* event. There exist at least three peculiar "statistical" signatures of dark matter (signal) events that can be used to discriminate against background events:

1. The periodic (seasonal) modulation due to the Earth motion around the Sun: this is a ~60 km/s effect, which translates into a ~10% signal "sinusoidal" modulation with a maximum in June and a minimum in December; for the first such estimate see Ref. [94].

Exercise 71. Convince yourself that in fact a ~60 km/s difference translates into a ~10% signal modulation.

2. Since the dark matter particle "wind" relative to the Earth is not aligned with the Earth's axis of rotation, there should also be a diurnal modulation due to eclipsing of the dark matter flux by the Earth (original calculation: Ref. [95]); as you can imagine, the amplitude of this effect depends on how strongly the dark matter particle interacts with nuclei; given what we (now) know about this, diurnal modulation should be quite small.

Exercise 72. Use the results in Table II of Ref. [95] to estimate what the maximal diurnal modulation amplitude is for a 10, 100, 1000 GeV particle (i.e. for particles with a scattering cross section set at the current best spin-independent limits at those masses).

3. Finally, the dark matter wind is aligned with the Sun's motion in the galaxy, so directional information can be used to discriminate against a background that, of all things, should certainly not care what the Sun does as it moves through the galaxy (for an early reference on this see Ref. [96]).

4.3 Direct detection event rates, for real

Suppose I give you a well-defined particle dark matter theory. How do you calculate the corresponding detector event rate R, i.e. the number of scattering events per unit time, energy and detector mass, for an event characterized by a recoil energy E_R? Here is how.

To begin with, the rate, as any rate in particle physics, is the product of the number of target nuclei[d] N_T, times the average flux of dark matter particles, which we can decompose in an average dark matter number density $n_\chi = \rho_{\rm DM}/m_\chi$ times a velocity v_χ (which we will need to appropriately average on), times a differential cross section $\mathrm{d}\sigma/\mathrm{d}E_R$; the latter cross section, in general, will depend on the dark matter velocity v_χ relative to the target material (we will see in detail specific cases below). In formulaic summary (and keeping track of the possibility that the cross section depends on velocity, thus bringing it inside the velocity average $\langle\cdot\rangle$):

$$\frac{\mathrm{d}R}{\mathrm{d}E_R} = N_T\, n_\chi \left\langle v_\chi \frac{\mathrm{d}\sigma}{\mathrm{d}E_R}\right\rangle. \tag{4.3}$$

Let us first tackle the kinematics. As often, it is easiest to go to the center-of-mass frame, where the dark matter particle initial (final, respectively) momentum $\vec{p}$ ($\vec{p}'$) and the target initial (final) momentum $\vec{k}_T$ ($\vec{k}'_T$) are equal and opposite:

$$\vec{p} = -\vec{k}_T = \mu_T \vec{v}_\chi, \tag{4.4}$$

$$\vec{p}' = -\vec{k}'_T = \vec{q} + \mu_T \vec{v}_\chi. \tag{4.5}$$

In the Eqs. (4.4) and (4.5) I have introduced the momentum transfer $\vec{q} = \vec{p}' - \vec{p}$, and I have indicated with μ_T the dark matter-nucleus reduced mass.

[d] Note that for sub-GeV dark matter scattering off of electrons in the target material might be more relevant than scattering off of nuclei; the relevant calculation is complicated by the fact that electrons are in bound states, but is otherwise straightforward, see e.g. [90,91].

For elastic scattering, in the center-of-mass frame, as the Reader recalls from her early days as a physicist, one has $|\vec{p}| = |\vec{p}'|$, and thus subtracting Eq. (4.5) from Eq. (4.4) and squaring one gets

$$\frac{q^2}{2} = p^2 - \vec{p} \cdot \vec{p}' = p^2(1 - \cos\theta) = \mu_T^2 v_\chi^2 (1 - \cos\theta),$$

with $\cos\theta$ the scattering angle in the center-of-mass frame. This allows us to finally derive the thing we would like to actually measure, i.e. the (kinetic) energy transferred from the dark matter particle to the nucleus (of mass m_T), also known as recoil energy:

$$E_R = \frac{q^2}{2m_T} = \frac{\mu_T^2}{m_T} v_\chi^2 (1 - \cos\theta).$$

Note that the *minimum* velocity at a given energy E_R (or momentum transfer) corresponds to backscattering (i.e. $\cos\theta = -1$), and is given by

$$v_{\rm min} = \sqrt{\frac{m_T E_R}{2\mu_T^2}} = \frac{q}{2\mu_T}.$$

The equation above has important implications. For instance, as expected, lighter dark matter candidates ($m_\chi \ll m_T$) have larger threshold velocities $v_{\rm min}$ for detection: this totally makes sense, as it takes a big velocity for a light dark matter particle to deposit enough energy E_R versus a heavier one! Inelastic processes further increase the minimal velocity, as the problem you solved above illustrated.

At a given velocity v, the differential recoil energy

$$dE_R = (d\cos\theta)(\mu_T^2/m_T)v^2. \tag{4.6}$$

Substitute now Eq. (4.6) into Eq. (4.3) now, and write the velocity average explicitly to finally get

$$\frac{dR}{dE_R} = N_T \frac{\rho_{\rm DM} m_T}{m_\chi \mu_T^2} \int_{v_{\rm min}}^{v_{\rm esc}} d^3v \frac{f(v)}{v} \frac{d\sigma}{d\cos\theta'}, \tag{4.7}$$

where $f(v)$ is your favorite dark matter halo velocity probability distribution function (see Sec. 1.7).

The formula above is completely general. The dark matter particle physics at this points feeds into the differential scattering cross section (and in the trivial kinematic factor $m_\chi \mu_T^2$, of course), which is the appropriate scattering amplitude squared summed over final and averaged over initial polarizations, as usual (see Appendix A.2). In the non-relativistic limit, it is customary to write the scattering matrix element $\mathcal{M}(q^2)$ as the Fourier transform of the WIMP-nucleus potential,

$$\mathcal{M}(q^2) \sim \int \langle f|V(\vec{r})|i\rangle e^{i\vec{q}\cdot\vec{r}} \mathrm{d}\vec{r},$$

with

$$V(\vec{r}) = \sum_{\text{nucleons } n} \left(G_s^n + G_a^n \vec{\sigma}_\chi \cdot \vec{\sigma}_n\right) \delta(\vec{r} - \vec{r}_n).$$

This is not a generic expression (for which the curious Reader is invited to consult e.g. Ref. [97]). Rather, in the expression above, the constants G_s^n and G_a^n are the effective four-fermion (or scalar–scalar– nucleon–nucleon, or vector–vector–nucleon–nucleon etc., as the case may be) coupling constants for the dark matter–nucleon interactions for scalar (or spin-independent, i.e. associated with scalar operators under three-dimensional rotations) and axial, or spin-dependent interactions. We will see below why these are usually the dominant terms in a generic dark matter–nucleon interaction.

Both terms are coherent as long as the wavelength associated with the momentum transfer (in natural units $1/q$) is much smaller than the size of the nucleus R_{nucleus}, i.e. if $qR_{\text{nucleus}} \ll 1$. At smaller wavelengths (or for larger nuclei) the dark matter scattering process starts probing the nuclear structure, and one needs to introduce *form factors* to account for the fact that the terms are now at least partly incoherent. In practice, this is usually done by introducing form factors $F(q^2)$ defined by

$$\mathcal{M}(q^2) = T(0)F(q^2).$$

Of course, form factors for the scalar and spin-dependent interactions are in general different, since mass and spin are not necessarily distributed in the same way in a nucleus, and they depend on the specific target nucleus.

Now, the task at hand is to calculate the amplitude $\mathcal{M}(q^2)$, and here is where the fun begins. This calculation is interesting, as it genuinely is a multi-layered process: first we need to find the effective dark matter-quark interaction in the non-relativistic limit (small q^2) — this is basically the same as going from the electroweak theory to the Fermi four-fermion theory (see Appendix A.3); second, we need to account for the quark content of the nucleons — protons and neutrons: we need to take the quark fields operators and express them in terms of nucleons; in practice, this means to evaluate "matrix elements" like $\langle n|\bar{q}q|n\rangle$, where n is a neutron or a proton; finally, we need a nuclear model to paste the dark matter–nucleon interaction onto the full-glory nucleus.

We will follow Ref. [98] and work out in detail the case of a heavy Dirac neutrino ν, for definiteness. Scattering with quarks proceeds through Z exchange. In the non-relativistic limit (i.e. for $q^2 \ll m_Z^2$, valid here), one "integrates out" the Z, and the amplitude reduces to Fermi's four-fermion interaction:

$$\sqrt{2}G_F \bar{\nu}(V_\nu - A_\nu\gamma_5)\gamma_\mu \nu\, \bar{q}(V_q - A_q\gamma_5)\gamma^\mu q,$$

with the standard $V - A$ assignments: $V_\nu = A_\nu + \frac{1}{2}$ and $A_q = T_{3q}$, $V_q = T_{3q} - 2e_q \sin^2\theta_W$, with T_{3q} the value of the third component of the weak isospin of the quark q, e_q its electric charge in units of the elementary electric charge e, and θ_W Weinberg's angle.

Now we ought to take the Dirac spinors and boil them down to the non-relativistic limit. It is an instructive exercise to write explicitly the Dirac spinors in terms of 2-component spinors in the standard way, and compute the four structures above.

Exercise 73. From the spinor solutions to Dirac's equation, calculate:

1. $\bar{u}(p')\gamma^0 u(p)$,
2. $\bar{u}(p')\gamma^i u(p)$,
3. $\bar{u}(p')\gamma^0\gamma^5 u(p)$,
4. $\bar{u}(p')\gamma^i\gamma^5 u(p)$,

where $i = 1, 2, 3$, and indicate which are the leading "pieces" in the non-relativistic limit of small momentum transfer. See Ref. [97] for the solution to this exercise.

The diligent Reader has by now convinced herself that the pieces that survive in the non-relativistic limit are the time-like component of the vector current $\bar{\nu}\gamma_0\nu \propto u^\dagger u$ and the space-like component of the axial current, $\bar{\nu}\vec{\gamma}\gamma_5\nu \propto u^\dagger\vec{\sigma}u$. This, incidentally, justifies the guess we used above for $\mathcal{M}(q^2)$.

Next task is to "sandwich" the contact interactions in a nucleon, and then in the nuclear state of interest. Let us start with the axial current. The relevant matrix elements are given by the spin content of the nucleon $|n\rangle$. The relevant coefficients are extracted from (polarized) deep-inelastic scattering data:

$$\langle n|\bar{q}\gamma_\mu\gamma_5 q|n\rangle = 2s_\mu^{(n)}\Delta q^{(n)},$$

where $s_\mu^{(n)}$ is the spin of nucleon n, a proton or a neutron (the two being related by an isospin rotation: e.g. $\Delta u^{(p)} = \Delta d^{(n)}$, $\Delta s^{(p)} = \Delta s^{(n)}$ etc.). The values for Δq are obtained from neutron and hyperon β-decays, from the naive quark model, from deep-inelastic scattering data, from $\nu p \to \nu p$ elastic scattering etc., and they are far from being accurately determined (see e.g. Refs. [98] and [100]).

Going back to the Dirac neutrino, the four-fermion coupling constants for the axial part

$$\sqrt{2} G_F A_\nu A_q \bar{\nu} \gamma_\mu \gamma_5 \nu \bar{q} \gamma^\mu \gamma_5 q,$$

then read, at the nucleon level, for the proton:

$$G_a^p = \sqrt{2} G_F A_\nu \left(A_u \Delta u^{(p)} + A_d \Delta d^{(p)} + A_s \Delta s^{(p)} \right),$$

and for the neutron

$$G_a^n = \sqrt{2} G_F A_\nu \left(A_u \Delta u^{(n)} + A_d \Delta d^{(n)} + A_s \Delta s^{(n)} \right).$$

The scalar interaction, as we saw above, arises from the time-like component of the vector current

$$\sqrt{2} G_F V_\nu V_q \bar{\nu} \gamma_\mu \nu \bar{q} \gamma^\mu q. \tag{4.8}$$

leading (see Exercise 74) to the four-fermion couplings:

$$G_s^p = \frac{G_F}{\sqrt{2}} \left(1 - 4 \sin \theta_W\right) V_\nu, \tag{4.9}$$

$$G_s^n = -\frac{G_F}{\sqrt{2}} V_\nu. \tag{4.10}$$

Exercise 74. Use vector current conservation to derive Eqs. (4.9) and (4.10) from Eq. (4.8).

It is now time to move up to the issue (layer number 3) of calculating the matrix elements of the nucleon spin operators in a *nuclear* state. At $q^2 = 0$ this is equivalent to calculating the average spin for neutrons $\langle S_n \rangle$ and for protons $\langle S_p \rangle$ in a given nucleus, while at non-zero momentum transfer there will be a form-factor suppression given by the specific nuclear wave functions. For example, the averaged amplitude-squared at $q^2 = 0$ for a nuclear spin $J \neq 0$ is

$$\overline{|\mathcal{M}(0)|^2} = \frac{4(J+1)}{J}\left|G_a^p\langle S_p\rangle + G_a^n\langle S_n\rangle\right|^2.$$

Exercise 75. Explain the spin pre-factor $4(J+1)/J$.

Note that for even–even nuclei, i.e. nuclei with even numbers of protons and neutrons, $J = 0$ and $\mathcal{M}(0) = 0$. For even–odd nuclei, with $J \neq 0$, we need to estimate the values for $\langle S_n\rangle$ and $\langle S_p\rangle$; there are a variety of different models,[e] with predictions differing by factors of 2 or so, but always with $\langle S_p\rangle \simeq 0$ and at most as large as a few percent of $\langle S_n\rangle$.

At $q^2 > 0$ nuclear spin form factors are needed; usually, the averaged matrix element squared is parameterized by isoscalar and isovector spin form factors $F^0_{\rm spin}(q^2)$ and $F^1_{\rm spin}(q^2)$ as:

$$\overline{|\mathcal{M}(q^2)|^2} = \frac{J+1}{J}\Big|(G_a^p + G_a^n)\langle S_p + S_n\rangle F^0_{\rm spin}(q^2)$$
$$+ (G_a^p - G_a^n)\langle S_p - S_n\rangle F^1_{\rm spin}(q^2)\Big|^2.$$

The spin form factors are derived in the context of complex models, and calculations well beyond the scope of the present discussion, but it is instructive to stare in the face of the following, useful approximate expressions:

$$F^0_{\rm spin}(q^2) \simeq \exp\left(-\frac{r_0^2 q^2}{\hbar^2}\right); \qquad F^1_{\rm spin}(q^2) \simeq \exp\left(-\frac{r_0^2 q^2}{\hbar^2} + i\frac{cq}{\hbar}\right),$$

with the constants r_0, $r_1 \approx 1$–2 fm, $c \simeq -0.3$–0.3 fm parameterizing individual nuclei [98].

For the spin-independent part, the averaged amplitude squared also requires a nuclear mass form factor $F_{\rm mass}(q^2)$, resulting in the expression

$$\overline{|\mathcal{M}(q^2)|^2} = \left|ZG_s^p + NG_s^n\right|^2\left|F_{\rm mass}(q^2)\right|^2,$$

with, of course, N and Z the number of neutrons and protons in the give nucleus. Given that neutron scattering experiments indicate that

[e]For example, the single-particle shell model [101], the odd-group model [102], interacting shell models [103].

$F_{\rm mass}(q^2) \approx F_{\rm e.m.}(q^2)$, the latter being the usual electromagnetic form factor, it is customary to use the Woods-Saxon form

$$F_{\rm mass}(q^2) \simeq \frac{3j_1(qR)}{qR} \exp\left(-\frac{(qs)^2}{2}\right),$$

with the electromagnetic radius $R \approx A^{1/3}$ fm, and a "surface thickness" $s \approx 1$ fm.

4.4 Taking it spin by spin

While the discussion above is instructive, often all of the nasty nuclear physics-y aspects are kindly pre-digested for us, and what we contend with are limits, or experimental sensitivities, in a plane defined by the dark matter particle mass and the spin-dependent or spin-independent interaction cross section with protons (or neutrons). This is illustrated for example in Fig. 4.1. Thus, in real life those limits or projected sensitivities are our input, and given a dark matter candidate, of a given mass and spin, we need to estimate those cross section. This is what I outline in this section.

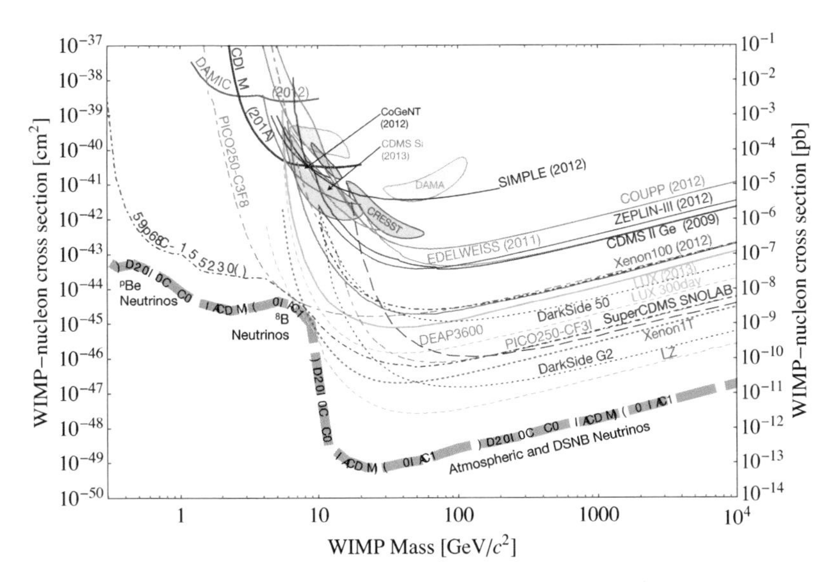

Figure 4.1: Overview of limits on dark matter spin-independent cross section off of nucleons, and projected sensitivities; the yellow-shaded region indicates the region where the diffuse neutrino background and the background of neutrinos from the Sun would dominate the signal (see Sec. 4.5). From Ref. [104].

Let us first consider a scalar dark matter particle ϕ, and suppose that we have been good enough to write down the effective interaction cross section with quarks at the relevant energy scale for direct detection (bear in mind that there are important renormalization group (RG) running effects if instead you consider the effective operator at some *higher* energy scale, as pointed out e.g. in Ref. [105]), something of the form

$$\frac{m_q}{\Lambda^2}\phi\phi\bar{q}(S_q + iP_q\gamma^5)q.$$

Note the m_q pre-factor: it is present for a variety of reasons: the interaction flips chirality, the m_q factor suppresses potentially troublesome flavor-changing interactions, and it is hard to write a UV-complete theory that does not come with that factor. As we discussed above, the pseudoscalar piece is velocity-suppressed, thus what we are left with is the scalar interaction.

It is customary to write an amplitude for scattering off of a nucleus (A, Z) (a proton will just be $(1, 1)$) as:

$$\mathcal{T} = \frac{2}{27}(S_c + S_b + S_t)Af^p_{Tg} + \sum_{q=u,d,s} S_q\left(Zf^p_{Tq} + (A - Z)f^n_{Tq}\right),$$

where the first term corresponds to the gluonic contributions from heavy quark loops, and the second term from the light valence quark contributions, i.e.

$$f^p_{Tq}m_p\langle p|p\rangle \equiv \langle p|m_q\bar{q}q|p\rangle, \qquad f_{Tg} \equiv 1 - \sum_{q=u,d,s} f_{Tq}.$$

Let us pause for a moment and appreciate what we have just done (in particular the seemingly mysterious 2/27 coefficient above), which actually contains some good lessons in particle physics.

Exercise 76. Heavy quarks contribute to the mass of the nucleon through the anomaly [106,107]: under the heavy-quark expansion, one can substitute in the nucleon matrix element [106]

$$m_Q\bar{Q}Q \rightarrow \frac{-2\alpha_s}{24\pi}GG,$$

(*Continued*)

Exercise 76. *(Continued)*

where $Q = c,\ b,\ t$. Show that the trace of the QCD energy-momentum tensor, θ^μ_μ, can then be cast as

$$\theta^\mu_\mu = m_u\bar{u}u + m_d\bar{d}d + m_s\bar{s}s + \sum_{Q=b,c,t} m_Q\bar{Q}Q - \frac{7\alpha_s}{8\pi}GG =$$

$$= m_u\bar{u}u + m_d\bar{d}d + m_s\bar{s}s - \frac{9\alpha_s}{8\pi}GG.$$

Now sandwich this expression on a nucleon state, $\langle n|\cdot|n\rangle$, and argue that

$$\langle n|m_Q\bar{Q}Q|n\rangle = \frac{2}{27}m_n\left(1 - \sum_{q=u,d,s} Tq\right) \equiv \frac{2}{27}m_n f_{Tg},$$

which defines the gluonic contributions through heavy quark loops mentioned above.

Putting everything together, the cross section off of a nucleus target is

$$d\sigma_{SI}^{\phi N} = \frac{2!}{4\pi\Lambda^4}\left(\frac{m_\phi m_N}{m_\phi + m_N}\right)^2\left(\frac{m_p}{m_\phi}\mathcal{T}\right)^2,$$

which for $N \equiv p$ gives a total cross section

$$\sigma_{SI} = \left(\frac{2m_p^2\mathcal{T}}{\Lambda^2(m_\phi + m_p)}\right)^2.$$

Naturally, there is no spin-dependent amplitude for scalar dark matter.

For Dirac fermions ψ, the generic effective operator, evaluated at the nuclear scale[f] looks like

$$\frac{1}{\Lambda^3}\bar{\psi}(S_\psi + i\gamma^5 P_\psi)\psi\ m_q\bar{q}(S_q + iP_q\gamma^5)q$$

$$+\frac{1}{\Lambda^2}\bar{\psi}\gamma_\mu(V + A\gamma^5)\psi\ \bar{q}\gamma^\mu(v_q + a_q\gamma^5)q.$$

[f] Let me take this opportunity to give one more reminder that couplings run!

The spin-independent part comes from two pieces: the vector–vector coupling Vv_q, and the scalar–scalar coupling $S_\psi S_q$, and the resulting scattering cross section off of protons reads

$$\sigma_{SI} = \frac{m_{\psi p}^2}{\pi \Lambda^4} \left|2Vv_u + Vv_d + S_\psi (m_p/\Lambda)\mathcal{T}\right|^2,$$

where we assumed valence quark model relations for the vector–vector part (for a nucleus (A, Z) you would replace $2 \to Z + A$ for the up-quark and $1 \to 2A - Z$ for the down quark coefficients), and where $m_{\psi p}$ is the dark matter-proton reduced mass. The SD cross section instead reads

$$\sigma_{SD} = \frac{4}{\pi} J(J+1) \frac{m_{\psi N}^2}{\Lambda^4} \left| \sum_{q=u,d,s} Aa_q \Delta q^{(p)} \right|^2 .$$

What about spin $\frac{1}{2}$ *Majorana* fermions? Due to long-lasting theoretical biases, I still find myself reviewing papers where "neutralinos" and "dark matter particles" are used inter-changeably. Yet, as disturbing as this is, neutralinos retain the status of "paradigmatic" dark matter particles,[g] and as such it is important to quickly look at how direct neutralino detection works. For a Majorana particle χ the vector and tensor currents identically vanish,

Exercise 77. Prove the statement above.

...and of the remaining currents the only pieces that survive in the non-relativistic limit are the time-like component of the scalar current, giving the spin-independent term, $\bar{\chi}\chi \propto \chi^\dagger\chi$, and the space-like axial current, giving the spin-dependent interaction $\bar{\chi}\vec{\gamma}\gamma_5\chi \propto \chi^\dagger\vec{\sigma}\chi$.

Axial currents arise in supersymmetry primarily from Z exchange, thus the story follows pretty much exactly what we discussed above for Dirac neutrinos, with the replacement $A_\nu \to A_\chi$, where the neutralino–neutralino-Z vertex is given by

$$-\frac{ig}{2\cos m\theta_W} A_\chi \gamma_\mu \gamma_5.$$

The resulting spin-dependent cross section is proportional to the so-called "Higgsino asymmetry",

$$\sigma_{SD} \propto |N_{13}^2 - N_{14}^2|^2.$$

[g] Which is also why I review them in Appendix A.4.

Scalar interactions, instead, arise e.g. from the exchange of scalar particles, say φ, so that at low momentum-exchange the $-1/m_\varphi^2$ propagator produces a four-fermion amplitude

$$-\frac{g_{\varphi\chi\chi}g_{\varphi qq}}{m_\varphi^2}\bar{\chi}\chi\bar{q}q.$$

Now, identifying

$$\frac{g_{\varphi\chi\chi}g_{\varphi qq}}{m_\varphi^2} \to \frac{1}{\Lambda^2},$$

we get a cross section, from our general discussion above, of

$$\sigma_{SI} = (2!)^2 \frac{m_{\chi p}^2}{\pi\Lambda^4}\left(\frac{m_p\mathcal{T}}{\Lambda}\right)^2.$$

Notice the $(2!)^2$ factor that comes from the two counts of Wick contraction for the Majorana field.

Note that both axial and scalar interactions of neutralinos with nucleons have additional contributions from s-channel scalar quark exchange. Such contributions are however typically (although not always!) subdominant, especially after LHC searches have put stringent constraints on how light scalar quarks can possibly be (see Appendix A.4).

Finally, for a spin-1 particle B^μ (e.g. in some incarnations the lightest Kaluza–Klein particle (LKP) of minimal universal extra-dimensions (UED), see Appendix A.5), one would write an effective Lagrangian

$$\frac{1}{\Lambda^2}B^\mu B_\mu S_q m_q \bar{q}q + \frac{A_q}{\Lambda^2} i\epsilon^{\alpha\mu\nu\beta}\left(B_\nu i \overleftrightarrow{\partial}_\alpha B_\mu\right)\bar{q}\gamma_\beta\gamma_5 q,$$

producing a spin-independent cross section entirely analogous to the scalar case,

$$\sigma_{SI} = \left(\frac{2m_p^2\mathcal{T}}{\Lambda^2(m_B + m_p)}\right)^2$$

and a spin-dependent cross section which, after a slightly more painful calculation and summation over polarizations, gives [108]

$$\sigma_{\rm SD} = (2!)^2 \frac{8}{3\pi} J(J+1)\frac{m_{BN}^2}{\Lambda^4}\Big|\sum_{q=u,d,s} A_q \Delta q^{(p)}\Big|^2.$$

There are other interesting possibilities. Let me make a couple of examples. First, suppose the dark matter interacts with SM particles via an effective "milli-charge", i.e. a charge εe, where e is the e of electromagnetism (thus the fine structure constant $\alpha = e^2/(4\pi)$, in natural units, and $\varepsilon \ll 1$), then the scattering cross section off a nucleus (A, Z) will be the same as ordinary Compton scattering of an electron (for example), rescaled by ε:

$$\sigma_N = \frac{16\pi\alpha^2\varepsilon^2 Z^2 \mu_N^2}{q^4},$$

with μ_N the dark matter-nucleus reduced mass. In this case, instead of e.g. the $1/q^2$ dependence we found for effective interactions of the form $g\bar{\chi}\chi\bar{q}q$, we have a $1/q^4$, and thus a totally different recoil spectrum! Similarly, if for some reason the coefficient of the effective scalar interaction $g \propto 1/q^2$, then again the cross section will have a dependence on momentum transfer $\propto 1/q^4 \sim 1/E_R^2$; if instead the leading effective operator is *proportional* to q, as is the case for the non-relativistic limits of an operator like $g\bar{\chi}\gamma^5\bar{q}q$, then $d\sigma/dE_R \sim E_R$.

A second example was suggested to me by Francesco D'Eramo,[h] together with the following very instructive exercise. It has to do with scattering off of nuclei of a dark matter model that does not, at tree level, interact with quarks at all!

Exercise 78. Consider a Dirac fermion dark matter particle χ coupled to a heavy vector mediator V through the interaction

$$\mathcal{L}_{\chi\chi V} = g_\chi V_\mu \overline{\chi}\gamma^\mu \chi\,.$$

Interactions with SM fields are mediated by V, which only couples to leptons (this is a "leptophilic" dark matter scenario):

$$\mathcal{L}_{llV} = g_l V_\mu \sum_{i=1}^{3} \overline{l^{(i)}}\gamma^\mu l^{(i)},$$

with the sum running over the three different SM generations (i.e. electron, muon, and τ). Spin-independent elastic scattering off of nuclei is mediated by the contact interaction

$$\mathcal{L}_{\rm DD} = -\frac{\overline{\chi}\gamma^\mu\chi}{m_V^2}\left[C_V^{(u)}\overline{u}\gamma^\mu u + C_V^{(d)}\overline{d}\gamma^\mu d\right],$$

(Continued)

[h]Who, as it turns out, is omniscient on this topic.

Exercise 78. (*Continued*)

where the dimension 6 interaction is normalized with the mediator mass m_V. As discussed above, the cross section for spin-independent scattering is

$$\sigma_{\rm SI} = \frac{1}{\pi\, m_V^4} \frac{m_\chi^2 m_{\mathcal{N}}^2}{(m_\chi + m_{\mathcal{N}})^2} \left| C_V^{(u)}(A+Z) + C_V^{(d)}(2A-Z) \right|^2 , \tag{4.11}$$

off of a target nucleus $\mathcal{N}$ with mass (atomic) number A (Z). Although, there is no tree-level contribution to spin-independent scattering in the leptophilic model, radiative corrections induce a dark matter coupling to light quarks. This effect is described by the RG equations [105,109]

$$\frac{d\, C_V^{(u)}}{d \ln \mu} = 3 \times \tfrac{8}{3} \frac{\alpha_{\rm em}}{4\pi} Q_u\, Q_l\, (g_\chi g_l)\,,$$

$$\frac{d\, C_V^{(d)}}{d \ln \mu} = 3 \times \frac{8}{3} \tfrac{\alpha_{\rm em}}{4\pi} Q_d\, Q_d\, (g_\chi g_l)\,.$$

where the overall factor of 3 accounts for the three leptons and Q_f is the charge of the SM fermion f.

(i) Solve the RG equations and show that

$$C_V^{(u)}(\mu) = -\frac{4\alpha_{\rm em}}{3\pi} (g_\chi g_l)\, \ln(\mu/m_V),$$

$$C_V^{(d)}(\mu) = \frac{2\alpha_{\rm em}}{3\pi} (g_\chi g_l)\, \ln(\mu/m_V).$$

(ii) Show that the resulting spin-independent cross section is given by

$$\sigma_{\rm SI} = \frac{1}{\pi\, m_V^4} \frac{m_\chi^2 m_{\mathcal{N}}^2}{(m_\chi + m_{\mathcal{N}})^2} (g_\chi g_l)^2 \frac{4\alpha_{\rm em}^2 Z^2}{\pi^2} \ln^2(\mu_N/m_V).$$

(iii) Indicate at which scale μ_N the couplings should be evaluated.

4.5 The neutrino floor

My esteemed colleague Joel Primack often reminds me that direct dark matter detection is a great business to be in (and one that he in fact contributed to get started, see Ref. [110]): if one *does not* see a signal, then one needs bigger, better detectors (to augment the experimental sensitivity); if one *does* see a signal, then one needs bigger, better detectors (to pinpoint and understand the dark matter particle properties). However, there is a limit to how far we can push the sensitivity of direct dark matter detection experiments: large-enough detectors are bound to start being sensitive to a well-known weakly interacting particle, the SM neutrino.

Extraterrestrial neutrinos have been conclusively detected from all of two sources: the Sun, and supernova SN1987A. We know the solar neutrino background quite well: most of the highest-energy solar neutrinos are produced in the decay $^8B \rightarrow {}^7B^* + e^+ + \nu_e$ (the so-called *hep* neutrinos are also cut-off at around 20 MeV and have much lower flux). The total flux of ^8B solar neutrinos, that we will use to make our estimates of the "irreducible" neutrino background is around 6×10^6 cm^{-2} s^{-1}. The "diffuse supernova neutrino background", or DSNB, is the accumulated flux of neutrinos from all supernovae, everywhere: you estimate it as an integral over the (core-collapse) supernova rate as a function of redshift times the (redshifted) neutrino spectrum per supernova. As you might imagine, there is some uncertainty in this estimate but, roughly, the flux is on the order of 10 cm^{-2} s^{-1}. Finally, neutrinos are produced in inelastic collisions of cosmic rays with nuclei in the atmosphere: these atmospheric neutrinos dominate the neutrino flux above (depending on estimates) 50 MeV, and the flux is approximately $\phi_{\rm atm} \sim 10^{-1}$ cm^{-2} s^{-1}.

Neutrino scattering off of nuclei is a well-known process, that pretty much works identically to what we described above: for a nucleus of "weak charge"

$$Q_W = N - (1 - 4\sin^2\theta_W)Z,$$

the cross section is

$$\frac{d\sigma}{dE_R} = \frac{G_F^2 Q_W^2 m_N^2}{4\pi}\left(1 - \frac{m_N E_R}{2E_\nu^2}\right) F(m_N E_R), \tag{4.12}$$

where the latter factor F is a form factor that accounts for the distribution of "weak charge" within the nucleus [111].

Exercise 79. Derive Eq. (4.12) using what you learned for scattering of a neutrino-like dark matter particle above.

Suppose we wanted to get a rough estimate of the dark matter–nucleon cross section, for a couple representative dark matter particle masses (say 5 and 100 GeV), such that the dark matter event rate were equal to the neutrino background event rate. This will give us an approximate ball-park for the neutrino floor. If we do things reasonably, we should get something similar to what shown with the thick, dashed orange line in Fig. 4.1.

First: what is the relevant neutrino background for the 5 and 100 GeV dark matter particles?

Exercise 80. Consider a nucleus of mass m_T =100 GeV.

(i) Show that the typical recoil energy from scattering of a 5 GeV dark matter candidate is around 0.25 keV, while it is of around 100 keV for scattering of a 100 GeV dark matter candidate;
(ii) Show that, approximately, the incident neutrino energy E_ν necessary to produce a (maximal) recoil energy E_R in the coherent scattering off of the same 100 GeV nucleus is $E_\nu \sim \sqrt{E_R m_T/2}$.
(iii) Show that the relevant neutrino energies to produce recoil events relevant for 5 GeV (100 GeV) dark matter scattering are $E_\nu \sim 2$ MeV ($E_\nu \sim 50$ MeV, respectively).

Exercise 80 indicates that the relevant neutrino background is solar neutrinos for light ($m_\chi \lesssim 10$ GeV) dark matter, and is DSNB or atmospheric neutrinos for more massive particles. The question now is: what is the dark matter–nucleon cross section for which scattering of neutrinos and dark matter give comparable rates?

Exercise 81. We will estimate the neutrino floor for light and heavy WIMPs.

(i) Estimate the "flux times cross section" for solar neutrinos, for $E_\nu \sim$ 2 MeV, using a flux at that energy of $\phi_\nu \sim 10^5$ cm^{-2} s^{-1}, and show that it is approximately $(\phi\,\sigma)_{\nu,\text{solar}} \sim 2 \times 10^{-38}$ s^{-1}.
(ii) Same for atmospheric neutrinos of 30 MeV energy, using a flux at that energy of $\phi_\nu \sim 10^{-2}$ cm^{-2} s^{-1}, and show that this time $(\phi\,\sigma)_{\nu,\text{atm}} \sim 3 \times 10^{-41}$ s^{-1}.

(Continued)

Exercise 81. *(Continued)*

(iii) Using default values for the dark matter density and velocity as in Eq. (4.2), estimate the cross section for which $(\phi\,\sigma)_\nu \sim (\phi\,\sigma)_{\rm DM}$ and show that it corresponds to

$$\sigma_{\nu-{\rm floor}}(5\ {\rm GeV}) \sim 10^{-44}\ {\rm cm}^2;$$
$$\sigma_{\nu-{\rm floor}}(5\ {\rm GeV}) \sim 4 \times 10^{-49}\ {\rm cm}^2,$$

in rough agreement with what shown with the thick, dashed orange line in Fig. 4.1.

Is there any way we can deal with the diffuse neutrino background once detectors reach that sensitivity? There exist a few possible avenues, for example using multiple target nuclei, measuring annual modulation, and directional information [112].

Getting down to the "neutrino floor'; sensitivity level will also be exciting from the standpoint of SM physics and high-energy astrophysics [113]. As many times in the history of particle physics, I am confident that my experimental colleagues will find a way, and prove wrong the notion that direct dark matter detection cannot move beyond the neutrino floor in just the same way as last century they proved wrong Bethe and Peierls' prediction that it was "absolutely impossible" to detect neutrinos.

Chapter 5

Indirect Dark Matter Searches

5.1 Praeludium

The fundamental (elementary) particle nature of dark matter can be probed *indirectly*, i.e. without directly detecting the particle itself, with the detection of the particle debris (photons, neutrinos, or charged cosmic rays) produced by, or affected by, dark matter as an elementary particle. The key processes are:

(a) the *pair-annihilation* of dark matter particles (which we shall generically indicate here with the symbol χ), producing Standard Model (SM) particles in the final state: $\chi + \chi \rightarrow$ SM,
(b) the *decay* of dark matter particles into SM particles: $\chi \rightarrow$ SM.

Other processes might exist, but are less common in the literature and in model building, and we will not entertain them here.[a] It is important to note that none of the processes listed above is bound to necessarily occur in any particle dark matter model: for example (b) does not occur if the dark matter is *absolutely* stable, and (a) can be highly suppressed, or, even, not occur at all, if the coupling of the dark matter sector to the SM is suppressed, or if the dark matter sector is somehow "secluded", or if, finally, the dark matter is not its own antiparticle, and there is no anti-dark matter around.

[a]Examples of processes I have had an interest in are "excitations" of dark matter by the interstellar thermal plasma [10, 114, 115] or by astrophysical jets [116], or, vice versa, absorption of light by dark matter [117].

There exist, however, reasons to be optimistic about the prospect of detecting non-gravitational signatures from dark matter via indirect detection: firstly, some of the best motivated (from a theoretical standpoint) extensions to the SM encompassing a dark matter candidate χ predict non-vanishing coupling of χ to SM particles; this would entail processes (a) or (b), at some level; secondly, there exist "phenomenological" reasons, chiefly motivated by the thermal relic paradigm I reviewed in Chapter 2. If indeed the dark matter particle is a thermal relic, indirect detection rates, from processes (a), are quantitatively connected with the observed abundance of dark matter. Important caveats of course exist, as I discussed in Chapter 3, such that the annihilation rate today might well be quite different from a naive estimate of the annihilation rate at the chemical decoupling temperature of the relic.

5.2 The ingredients of indirect dark matter searches

There are three key ingredients to understand indirect dark matter detection at a qualitative level, and to be able to make quantitative predictions:

1. *Production rates* of the relevant SM particles (sometimes also known as "messengers", but not in the model-building sense of mediators of interactions, beware!); this is related to the pair-annihilation or decay rate of the dark matter particle;
2. *Energy scale* of the SM messengers: this is set by the mass of the dark matter particle[b];
3. *Annihilation products*: this largely model-dependent ingredient specifies which SM particles are produced by the event.

The rate $\Gamma_{e^{\pm},\bar{p},\gamma,\nu,\ldots}$ for a given SM messenger (i.e. the flux of a given SM particle species per unit time from a unit volume V containing dark matter particles) is generically the product of three factors: (1) the number of dark matter particle pairs (or of dark matter particles, for decaying χ) in the volume V times (2) the pair-annihilation (or decay) rate,

[b]There are exceptions to this statement, for example from decay chains kinematics.

times (3) the number of given SM particles per annihilation (decay) event. In formulae[c]:

$$\Gamma_{\text{SM,ann}} \sim \left(\int_V \frac{\rho_{\text{DM}}^2}{m_\chi^2} \mathrm{d}V\right) \times (\sigma v) \times \left(N_{\text{SM, ann}}\right),$$

$$\Gamma_{\text{SM,dec}} \sim \left(\int_V \frac{\rho_{\text{DM}}}{m_\chi} \mathrm{d}V\right) \times \left(\frac{1}{\tau_{\text{dec}}}\right) \times \left(N_{\text{SM,dec}}\right).$$

Exercise 83. Verify that $\Gamma_{\text{SM, ann}}$ and $\Gamma_{\text{SM,dec}}$ have the correct dimensions of rate (inverse time).

It is interesting for the present discussion that many of the key quantities (ρ_{DM}, m_χ, $\sigma v, \ldots$) are potentially connected to how the dark matter was produced in the early universe. The dark matter production mechanism in the very early universe is therefore a great starting point both for model building, for estimating the relevant indirect detection techniques, and for setting constraints.

What have we learned thus far about annihilation processes that we could use to detect non-gravitational signals from particle dark matter?

— $\langle \sigma v \rangle \sim 3 \times 10^{-26}$ cm^3/s is a good "miraculous" benchmark,
— the miraculous number above is not weakly interacting massive particles (WIMPs)-specific, and is independent of mass, up to logarithmic effects,
— there exist numerous caveats, both from the particle physics side (co-annihilation, resonances, thresholds, etc.) and from the cosmology side (quintessence, non-thermal production, asymmetry, etc.).

[c]Note that if the dark matter is its own antiparticle, a factor 1/2 should appear for the annihilation formula, since the number of pairs given N particles in the volume V is $\sim N^2/2$ for large N. Similarly, if the dark matter is half composed of particles half of antiparticles, but ρ_{DM} indicates the total dark matter density, then there should be a factor 1/4. It is left as Exercise 82 to calculate the pre-factor for a scenario with asymmetric dark matter.

What about dark matter decay? If dark matter is unstable with a lifetime well in excess of the age of the universe, the decay products would also be a great way to detect non-gravitational signals from dark matter! From a theoretical standpoint, a GUT-scale or Planck-scale dark matter number-violating operators should be generic (see e.g. my recent paper on this topic [118]). For example for a dimension-5 operator,

$$\Gamma_5 \sim \frac{1}{M^2} m_\chi^3,$$

the resulting lifetime

$$\tau_5 \sim 1 \text{ s} \left(\frac{1 \text{ TeV}}{m_\chi}\right)^3 \left(\frac{M}{10^{16} \text{ GeV}}\right)^2, \tag{5.1}$$

would potentially result in an impact on Big Bang Nucleosynthesis (BBN), but it would imply an excessively short-lived dark matter candidate.

Exercise 84. Make sure you refresh your memory about "natural units" and how to convert energies to times, and convince yourself of the overall factor of 1 s in Eq. (5.1).

For a dimension-6 operator things look more interesting:

$$\Gamma_6 \sim \frac{1}{M^4} m_\chi^5,$$

with a lifetime

$$\tau_6 \sim 10^{27} \text{ s} \left(\frac{1 \text{ TeV}}{m_\chi}\right)^5 \left(\frac{M}{10^{16} \text{ GeV}}\right)^4,$$

which turns out to be a very interesting lifetime range to explain, for example, the Pamela/AMS-02 positron excess [119,120] and for searches for dark matter with gamma rays, as we shall see later on.

The particle mass range for dark matter decay is entirely open. In Chapter 9 we will see instances of potentially unstable dark matter candidates with very large masses; axions (Chapter 7) and sterile neutrinos (Chapter 8) have typical masses in the μeV and keV, respectively, and are both predicted to decay on cosmologically and astronomically (possibly) interesting timescales (i.e. giving rise to signals that for some model realizations could be detectable).

Exercise 85. Consider a cluster of galaxies of mass $10^{15}\ M_{\odot}$ 100 Mpc away. Assume that your dark matter particle candidate decays to two monochromatic photons. Find the lifetime of a dark matter particle with a mass anywhere between a nano-eV and a Tera-eV that would produce a flux of photons from that cluster of the same order of magnitude as the extragalactic background light at the frequency corresponding to the photons from the dark matter particle decay. For an estimate of the extragalactic background light intensity see e.g. Ref. [121].

For definiteness, let me now concentrate on dark matter annihilation. Let us now try to corner the key ingredients to make predictions for indirect searches for dark matter. First and foremost: what do we know about the dark matter particle mass, which sets the energy scale for the particles produced in an annihilation event? We saw earlier that for the specific case of WIMPs a useful lower limit is provided by the Lee–Weinberg bound at about 10 GeV,[d] while unitarity constrains WIMPs, on the large mass end, to be lighter than a few 100 TeV. I think this is a reasonable range to keep in mind, if one is wed to the notion of a weakly interacting dark matter particle.

There exist, however, a number of theoretical prejudices that have populated this field for a long time. A historically interesting one has it that WIMPs must be heavier than 40 GeV. This prejudice somehow even managed to distort how certain dark matter search experiments were optimized! It is worthwhile then to see where this prejudice comes from.

The first tenet of the "WIMPs must be heavier than 40 GeV" prejudice is that WIMPs are supersymmetric neutralinos. Browsing papers on dark matter (especially on *astro-ph*) the confusion between WIMPs and neutralinos is not unheard of. The second tenet is that there exists one universal soft supersymmetry breaking mass for all three gauginos at the GUT scale. If $M_1(M_{\rm GUT}) = M_2(M_{\rm GUT})$, where M_1 is the soft supersymmetry breaking scale associated with the $U(1)_Y$ gaugino and M_2 that of the SU(2) gaugino, renormalization group (RG) evolution (and the assumption of a "desert" between the electroweak and the GUT scale) implies

[d] Although of course such light "WIMPs" are constrained by the invisible decay width of the Z and thus some model-building massaging is inevitably in order.

that $M_1(M_{\rm EW}) \simeq 0.4 \times M_2(M_{\rm EW})$. Now, LEP2 constrains the chargino mass to be above about half its center-of-mass energy, or $m_{\tilde{\chi}_1^\pm} \gtrsim 100$ GeV. But one of the charginos (the wino-like, in SUSY slang) has a mass very close to $M_2(M_{\rm EW})$, therefore implying that the lightest, bino-like neutralino $m_{\tilde{\chi}_1^0} \simeq M_1(M_{\rm EW}) \gtrsim 0.4 \times 100$ GeV $= 40$ GeV. Amazing. Of course, GUT-scale universality, RG evolution etc. are all model-dependent ingredient, not to mention the assumption that the dark matter is a neutralino.[e]

What do we know about the annihilation final state? *Presque rien*, almost nothing. If the dark matter particle is a Majorana fermion, then the pair-annihilation into a fermion–antifermion final state is "helicity suppressed": $\chi\chi \to f\bar{f}$ requires a helicity flip, and thus the matrix element squared is proportional to the square of the fermion mass, $|M|^2 \propto m_f^2$ (in just the same way as for charged pion decay — a Majorana pair in a $l = 0$ wave is in a *CP*-odd state). As a result, pair-annihilation of Majorana dark matter into light fermions is highly suppressed. If $m_\chi < m_{\rm top}$ and if the annihilation channel $\chi\chi \to$ bosons (such as W^+W^- or hh) is suppressed, the dominant annihilation final states are $b\bar{b}$ and $\tau^+\tau^-$. This explains the otherwise surprising popularity of these two final states in the literature on dark matter indirect detection.[f] Note that besides $m_b \gg m_\tau$, the bottom quark final state wins by an additional factor 3 from color. There exist, however, circumstances where the $\tau\tau$ final state can be boosted, for example with a light scalar tau in supersymmetry (in the so-called stau co-annihilation region). Of course, leptophilic models with couplings proportional to lepton mass would also predict a dominantly $\tau\tau$ annihilation final state.

In universal extra dimension (UED) (see Appendix A.5) the situation is entirely different: the particle that is usually the stable lightest Kaluza–Klein excitation is the $n=1$ mode of the hyper-charge gauge

[e]The feedback I received on this paragraph from Francesco D'Eramo was especially interesting: he points out that this paragraph is "a bit anachronistic" as in recent years WIMP phenomenology has revolved around simplified models or effective field theories, and not supersymmetry; It made me realize that I belong to the "old" generation, where dark matter phenomenology was done with supersymmetric neutralinos. I still consider that an important case of interest. Yet, Francesco's comment most definitely made me feel old.

[f]If you are a graduate student at UCSC who happened to work on dark matter, and I sit on your thesis committee, I will ask you the reason why $b\bar{b}$ is such a popular final state during your thesis defense. Be ready.

boson, or $B^{(1)}$. The matrix element squared for pair-annihilation into a fermion–antifermion pair is proportional to the fourth power of the fermion's hyper-charge, $|M|^2 \propto |Y_f|^4$, thus up-type quarks ($Y_{u_L} = 4/3$) and charged leptons ($Y_{e_R} = 2$) are the preferred annihilation modes.

If the dark matter lives in an SU(2) multiplet (for example, higgsinos and winos in supersymmetry) everything is fixed by gauge interactions. For wino-like dark matter, the preferred final state is W^+W^-. The lightest neutralino is quasi-degenerate with the lightest chargino, with mass splittings on the order of a fraction of a GeV, and both lie at a scale close to M_2, the corresponding soft supersymmetry breaking mass. Co-annihilation plays obviously a very significant role, and co-annihilation is, here, of the "symbiotic" type (charginos pair-annihilate quite efficiently). The resulting pair-annihilation cross section is, approximately (and at tree level) [122]

$$\langle \sigma v \rangle_{\tilde{W}} \simeq \frac{3g^4}{16\pi M_2^2},$$

and the resulting thermal relic abundance is

$$\Omega_{\tilde{W}} h^2 \simeq 0.1 \left(\frac{M_2}{2.2\ \mathrm{TeV}} \right)^2,$$

implying that thermal winos must weigh about 2.2 TeV (again, this is a *tree-level* result: as it turns out Sommerfeld enhancement effects are quite important in this case, see e.g. Ref. [123], pushing the favored mass for thermal winos up to 2.8 TeV). Higgsinos, instead, come in a set of two neutralinos and a chargino with again small mass splittings and at a mass scale around μ. The dominant annihilation final states are W^+W^- and ZZ, and the pair-annihilation and thermal relic densities are given by

$$\langle \sigma v \rangle_{\tilde{H}} \simeq \frac{g^4}{512\,\pi\,\mu^2} \left(21 + 3\tan^2\theta_W + 11\tan^4\theta_W\right)$$

and

$$\Omega_{\tilde{H}} h^2 \simeq 0.1 \left(\frac{\mu}{1\ \mathrm{TeV}} \right)^2,$$

indicating that thermal higgsinos like to weigh about a TeV (again, at tree level). Note that while co-annihilation are important to determine the mass relevant for thermal relic production of the right amount of dark matter in the universe, it does not contribute to the pair-annihilation of dark matter particles today. The relevant late-universe cross section involves exclusively dark matter particles.

5.3 Indirect detection: warm-up lap

Time to charge ahead on indirect detection. The name of the game is to get "enough" number counts. In symbols,

$$N = \phi_\chi \cdot A_{\rm eff} \cdot t_{\rm exp},$$

where:

- ϕ_χ indicates the relevant dark matter-induced event rate, such as the flux of a certain type of SM particle in the appropriate angular region, and has units $\mathrm{cm}^{-2} \cdot \mathrm{s}^{-1}$.
- $A_{\rm eff}$ is an effective area: good to have in mind some numbers here. For example, in the business of gamma-ray telescopes, the Fermi Large Area Telescope (LAT) has an effective area of about 1 m^2, while the top-of-the-line atmospheric Cherenkov (ground based) telescopes, such as H.E.S.S. or MAGIC or VERITAS have effective areas on the order of 10^5 m^2 (and the future Cherenkov Telescope Array (CTA) one to two orders of magnitude larger!); the relevant numbers for the two key antimatter satellites are ~ 0.01 m^2 for Pamela and ~ 0.1 m^2 for AMS-02; finally, if you are asked to quote a number for high-energy neutrino telescopes, mention IceCube, and mumble 1 km^2. Note that if the flux is given per unit solid angle, the relevant detector quantity is the *acceptance*, with units of area times angle. For instance, the LAT has an acceptance of about 1.5–2.5 m^2 sr, depending on energy and direction.
- $t_{\rm exp}$ indicates the relevant "exposure time": for satellites this is on the order of a year, which as you know equals $\pi \times 10^7$ s; for typical ground-based telescopes you can perhaps count on about 100h, or about 10^5 s, while balloon experiments have a typical exposure time of the order of a week, or about 10^6 s.

To detect a signal we need to fulfill two basic conditions:

(i) have some *signal events*, i.e. $\phi_\chi \cdot A_{\rm eff} \cdot t_{\rm exp} \gg 1$,
(ii) have enough *signal-to-noise*, for example requesting $N_{\rm signal} > (\#\sigma)\sqrt{N_{\rm background}}$.

I like to classify astrophysical probes of dark matter into three categories:

1. **Very indirect**: this category includes effects induced by dark matter on astrophysical objects (**1.a**) or on cosmological observations (**1.b**);
2. **Indirect**: I include in this category probes that do not "trace back" to the annihilation event, as their trajectories are bent as the particles propagate: charged cosmic rays (Sec. 5.4);
3. **Not-so-indirect**: neutrinos (Sec. 5.5) and gamma rays (Sec. 5.6), with the great added advantage of traveling in straight lines (up to general relativistic corrections, that is...).

Let me say just a few words on items 1.a and 1.b, and some general remarks on 2., before delving in a somewhat more detailed discussion about cosmic rays, neutrinos, and gamma rays.

1.a Effects on Astrophysical Objects: folks have thought about an amazing variety of possibilities, including:

- *Solar Physics* (dark matter can affect the Sun's core temperature, the sound speed inside the Sun,...).
- *Neutron Star Capture*, possibly leading to the formation of black holes (BHs) (notably e.g. in the context of asymmetric dark matter, see e.g. [124]).
- *Supernova and Star cooling* (see the excellent book by Georg Raffelt [125]; we will talk about this in the chapter on axions, 7).
- *Protostars* (e.g. WIMP-fueled population-III stars, available also in Swedish [126]).
- *Planets warming* (the earliest paper I know on the topic is Ref. [127]).

Given the relevance of global warming to the general public (and to funding agencies), let us make an estimate of this latter effect. The "capture probability" for WIMPs is roughly

$$n_{\rm nucleons} \cdot \sigma_{\chi-N} \cdot R_{\rm planet} \lesssim 10^{-4},$$

where for the right-hand side I have used $n_{\rm nucleons} \sim N_A/{\rm cm}^3$, the current upper limit on the spin-dependent WIMP-nucleon cross section $\sigma_{\chi-N} \lesssim 10^{-37}\ {\rm cm}^2$ and the radius of Uranus, $R \sim 3 \times 10^9$ cm — the choice of Uranus is motivated by an anomalous heat observed in the planet, of about 10^{14} W.

Now, the power produced by dark matter assuming that all of the dark matter mass is converted to heat is

$$W \sim (\text{capture probability}) \cdot \pi R^2_{\text{planet}} \cdot \rho_{\text{DM}} \cdot v_{\text{DM}} \lesssim 10^{12}\ \text{W},$$

which tells us that we fall short by a couple orders of magnitude of explaining Uranus' anomalous heat. Too bad.

Exercise 86. Estimate the heat produced by dark matter annihilation in the Earth and compare with the accuracy of geothermal models (see also the much, much more refined discussion in [128]); how large should the dark matter–nucleon scattering cross section to cause global warming concerns?

1.b: Effects on Cosmology: lots of work here, which I do not intend to review here, spanning from effects on BBN, on the cosmic microwave background (CMB), on reionization, on structure formation and many more. Too many to even only give you a sensible laundry list! Go browse the arXiv and have fun!

5.4 Charged cosmic rays

Do we expect enough cosmic rays from dark matter annihilation or decay to detect a signal over the background? The ballpark energy density of cosmic rays in the Milky Way is

$$\epsilon_{\text{CR}} \sim 1\ \frac{\text{eV}}{\text{cm}^3}.$$

Let us estimate the energy density in cosmic rays dumped by dark matter annihilation in the galaxy:

$$\epsilon_{\text{DM}} \sim m_\chi \cdot \langle\sigma v\rangle \cdot n^2_{\text{DM}} \cdot t_{\text{MW}}.$$

To plug in numbers, with $m_\chi \sim 100$ GeV, $\rho_{\text{DM}} \sim 0.3$ GeV/cm^3, $\langle\sigma v\rangle \sim 3 \times 10^{-26}$ cm^3/s, and the Milky Way age $t_{\text{MW}} \sim 10 \times 10^9$ yr, I get

$$\epsilon_{\text{DM}} \sim 10^{-2}\ \frac{\text{eV}}{\text{cm}^3}.$$

Exercise 87. Improve on the estimate above using a Navarro–Frenk–White dark matter density profile and integrating over an appropriate cosmic-ray "diffusion region", e.g. a cylindrical slab of half-height 1 kpc and radius 20 kpc.

Exercise 88. Same as the exercise above, but for a decaying dark matter particle, find $\epsilon_{\rm DM}(\tau)$ where τ is the dark matter lifetime. Do you expect to get interesting limits on τ from this calculation? If yes, please mention me and this book in the acknowledgements of your forthcoming paper!

The estimate above indicates that the contribution of annihilating dark matter to cosmic rays is, at best, sub-dominant to the observed cosmic-ray energy density, but that it could be an $\mathcal{O}(1\%)$ effect. In fact, models of Galactic cosmic rays decently match observation, so this is in some sense good news for dark matter model building! As a result, it is key in this business to target *under-abundant* species, namely either heavy nuclei or antimatter (for example positrons (e^+), antiprotons ($\bar{p}$), antideuterons $\bar{D}$,...). Unfortunately, it is quite hard to produce heavy nuclei from dark matter annihilation (an event that produces, for example, a couple of high-energy jets only if the dark matter pair-annihilates to a pair of strongly-interacting particles). Antimatter, on the other hand, is promising; typical dark matter models (exceptions are certain flavors of asymmetric dark matter) are democratic in producing as much matter as antimatter in the annihilation or decay final products.

Figure 5.1 illustrates the *differential yield* (number of particles dN in a given energy interval, in units of the dark matter particle mass in the figure) of several particle species resulting from $\chi\chi \to q\bar{q}$, with q a light quark (left), and from $\chi\chi \to W^+W^-$ (I took these two nice figures from Ref. [129]; you can produce your own with their code, http://www.marcocirelli.net/PPPC4DMID.html). The red lines indicate photons, the black lines neutrinos, while the green and blue lines indicate $e^\pm$ and $\bar{p}$, respectively. All of these particle species primarily originate from the hadronization and cascade decays of jets initiated by the final state q and $\bar{q}$, or directly from the prompt decay modes of the W (notice the green

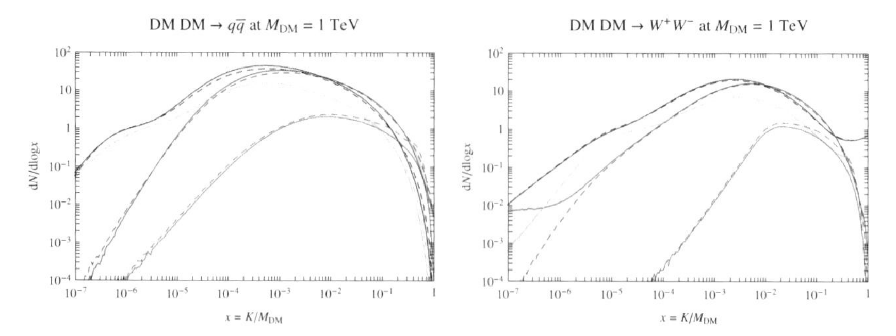

Figure 5.1: The differential photon (red lines), neutrino (black lines), $e^{\pm}$ (green lines), $\bar{p}$ (blue lines) yield from dark matter pair-annihilation into a (light) $q\bar{q}$ pair (left) and W^+W^- (right).
Source: Adapted from Ref. [129].

and black lines getting "horizontal" at $x = 1$, where x is the particles' kinetic energy normalized by the dark matter mass). The shaded region is a proxy for systematic theory uncertainties — the bracketing lines come from two different parton shower algorithms.

In cosmic rays, antimatter is primarily produced by spallation processes, such as

$$p + p \to p + p + \bar{p} + p,$$

where one of the protons in the initial state is a high-energy particle, and the second one is typically an H^+ nucleus in the interstellar medium gas, and baryon number conservation forces you to produce at least four nucleons in the final state. The process has a relatively large threshold (as a special relativity refresher carry out the two-lines calculation, Exercise 89), $E_p \gtrsim 7$ GeV. Now, the spectrum of cosmic rays observed in the galaxy falls steeply with energy,

$$\frac{dN_{\text{cosmic-ray protons}}}{dE} \sim E^{-2.7},$$

so compared to the maximal flux of cosmic-ray protons, observed at $E \sim 0.1$ GeV, antiprotons will be under abundant, at 0.1 GeV, by about a factor

$$\frac{\bar{p}}{p} \sim \left(\frac{0.1}{7.5}\right)^{2.7} \sim 10^{-5}.$$

This is in fact in remarkable agreement with what is observed, see Fig. 5.2, left, from Ref. [130].

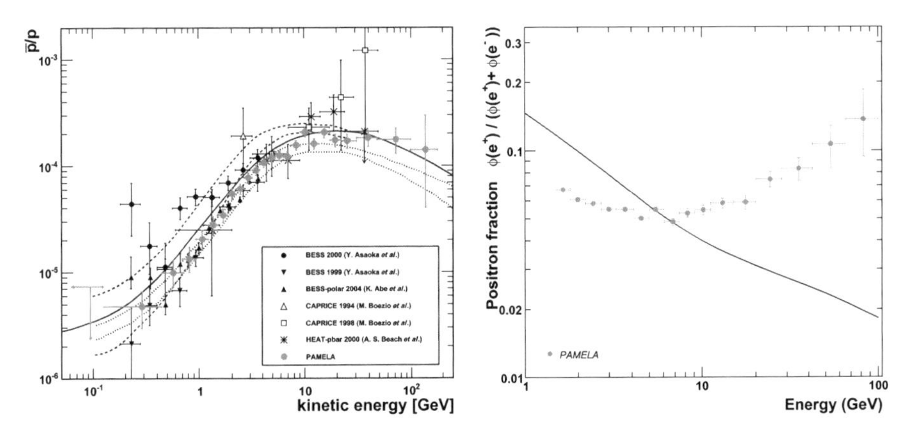

Figure 5.2: The cosmic-ray antiproton to proton ratio (left, from Ref. [130]) and the positron fraction (right, from Ref. [119]) as measured by the Pamela experiment.

There are therefore two effects that make antiprotons an interesting probe of dark matter (that, as Fig. 5.1 shows, tends to produce low-energy antinucleons): on the one hand there are few "beam" particles to produce cosmic-ray antiprotons, since the cosmic-ray proton spectrum falls steeply, and on the other hand the typical kinetic energy inherited by the final state antiproton will be of the same order as the threshold for the process. Indeed, Fig. 5.2, left, shows that the $\bar{p}/p$ ratio peaks right around 10 GeV, a much higher energy than the typical anti nucleon produced by dark matter. These two effects are even more drastic for anti-deuterons (i.e. bound states of $\bar{p}$ and $\bar{n}$), for which the key astrophysical background comes from the reaction

$$\bar{D} : p + p \to p + p + \bar{p} + p + \bar{n} + n,$$

that has a threshold of about 17 GeV (carry out the second little relativity one-liner calculation, Exercise 90). In addition, $\bar{D}$ have a hard time loosing energy by elastic scattering (tertiary population) since the deuteron binding energy is very low, and when hit $\bar{D}$ tend to disintegrate rather than lose energy! There is a smart idea out there (the proposed apparatus is called GAPS[g] [131]) to target specifically low-energy anti-deteurons and to detect them via the peculiar de-excitation X-rays that an atom capturing a $\bar{D}$ would produce.

[g]General Anti-Particle Spectrometer.

Exercise 91. What is the energy spectrum of a hydrogen atom with the electron replaced by a $\bar{D}$?

The key complication in using charged cosmic rays to probe dark matter is that the dark matter "source term" is tangled with the effects of propagation and energy losses as the produced particles make their way to our human detectors. How do we model cosmic-ray transport? The most successful framework is provided by the so-called diffusion models (adequate for cosmic-ray energies $E_{\rm CR} \lesssim 10^{17}$ eV). Let us indicate the differential (in energy) number density of cosmic rays with

$$\frac{dn}{dE} = \psi\,(\vec{x}, E, t)\,.$$

The master equation of cosmic-ray diffusion models looks something like this:

$$\frac{\partial}{\partial t}\psi \;=\; D(E)\Delta\psi \;+\; \frac{\partial}{\partial E}\,(b(E)\,\psi) \;+\; Q\,(\vec{x}, E, t), \tag{5.2}$$

The first term on the right-hand side describes diffusion, the second one energy losses, and the third includes all possible sources. As always in life and in science, it is possible (and relatively easy!) to add complications — an incomplete list of popular ones and of the associated recipes is:

- *Cosmic-ray convection*; recipe: add: $\frac{\partial}{\partial z}(v_c \cdot \psi)$;
- *Diffusive reacceleration*; recipe: add: $\frac{\partial}{\partial p}\, p^2\, D_{pp} \frac{\partial}{\partial p} \frac{1}{p^2}\, \psi$;
- *Fragmentation and decays*; recipe: add: $-\frac{1}{\tau_{\rm f,d}}\,\psi$.

When dealing with partial differential equations, we all learned in kindergarten that it is crucial to define *boundary conditions*. A popular choice is free-escape at the boundaries of a "diffusive region", whose geometry, for obvious reasons in the case of our own galaxy, is typically chosen to be a cylindrical slab, with

$$R \sim \mathcal{O}(1) \times 10\ \text{kpc},$$
$$h \sim \mathcal{O}(1) \times 1\ \text{kpc}.$$

These numbers (very) approximately reflect the distribution of gas and stars in the Milky Way.

The diffusion coefficient (that in certain models can depend also on position — it more than likely does in reality!) has a dependence on energy (a remnant of the fact that the Larmor radius scales with the particle's momentum!) that can be schematically cast as

$$D(E) \sim D_0 \left(\frac{E}{E_0} \right)^{\delta}, \qquad E_0 \sim \text{GeV}, \; D_0 \sim \text{few} \times 10^{28} \, \frac{\text{cm}^2}{\text{s}}, \qquad \delta \sim 0.7.$$

The parameters entering cosmic-ray diffusion are tuned self-consistently to reproduce key observational data, such as stable pure secondary to primary ratios as a function of energy (classic example: boron to carbon, B/C) or unstable secondary to primary ratios, such as $^{10}\text{Be}/^{9}\text{Be}$. For example, this latter ratio constrains quite severely the height of the diffusion region.

What are the relevant timescales for the diffusion equations? Two key quantities are the diffusion and the energy loss time scales:

$$\tau_{\text{diff}} \sim \frac{R^2}{D_0} \cdot E^{-\delta}, \qquad \tau_{\text{loss}} \sim \frac{E}{b(E)},$$

where R is the linear size of the diffusion region, or the relevant time/distance scale for which we want to calculate the typical associated diffusion length (for example, to infer which diffusion length corresponds to the energy loss time scale, we would plug in $R \sim c/\tau_{\text{loss}}$). Now bear with me[h] and try to simplify the partial differential equation in Eq. (5.2). On dimensional grounds, let us perform the following substitutions:

$$\Delta \to \frac{1}{R^2}, \qquad \frac{\partial}{\partial E} \to \frac{1}{E}.$$

The steady-state version of the diffusion Eq. (5.2) then becomes simply

$$0 = -\frac{\psi}{\tau_{\text{diff}}} - \frac{\psi}{\tau_{\text{loss}}} + Q,$$

implying that

$$\psi \sim Q \cdot \min[\tau_{\text{diff}}, \tau_{\text{loss}}]. \tag{5.3}$$

---

[h]...and remember we are *physicists*, not *mathematicians*, after all...

Let us see if this makes any sense, and consider cosmic-ray protons and primary and secondary electrons and positrons:

- If the primary sources of cosmic-ray protons are supernova remnants, and if the injected particles are accelerated via a Fermi first-order acceleration mechanism, we expect

$$Q \sim E^{-p}, \quad p \simeq 2.$$

Energy losses for protons in the GeV–TeV range are relatively inefficient, and typically $\tau_{\rm diff} \ll \tau_{\rm loss}$, therefore Eq. (5.3) would predict

$$\psi \sim E^{-2} \cdot E^{-\delta} \sim E^{-2.7},$$

which is in great agreement with observation!

- For primary electrons, let us suppose that again $Q \sim E^{-2}$, for example because the acceleration site is the same as for cosmic-ray protons (not such an unreasonable assumption). At high energy ($E_e \gg$ GeV) the dominant energy loss mechanisms are inverse-Compton scattering (i.e. the process of a high-energy electron up-scattering an ambient photon — the inverse of the classic Compton scattering where a high-energy photon up-scatters an electron at rest!) and synchrotron. Both have the same dependence on energy, $\propto E^2$, and the resulting energy loss term reads

$$b_e(E) \simeq b^0_{\rm IC}\left(\frac{u_{\rm ph}}{1\ {\rm eV/cm^3}}\right) \cdot E^2 + b^0_{\rm sync}\left(\frac{B}{1\ \mu{\rm G}}\right)^2 \cdot E^2,$$

where, in units of 10^{-16} GeV/s, the constants

$$b^0_{\rm IC} \simeq 0.76, \qquad b^0_{\rm sync} \simeq 0.025$$

and where $u_{\rm ph}$ corresponds to the background radiation energy density and B to the ambient magnetic field. Depending on the size and geometry of the diffusion region, Eq. (5.3) predicts a break between a low-energy regime where $\tau_{\rm diff} \ll \tau_{\rm loss}$, and

$$\psi_{\rm primary,\ low\text{-}energy} \sim Q \cdot \tau_{\rm diff} \sim E^{-2} \cdot E^{-\delta} \sim E^{-2.7},$$

and a high-energy regime where $\tau_{\rm diff} \gg \tau_{\rm loss}$, and

$$\psi_{\rm primary,\ high\text{-}energy} \sim Q \cdot \tau_{\rm loss} \sim E^{-1} \cdot \frac{E}{E^2} \sim E^{-3}.$$

The general prediction is thus of a *broken power-law* with a break corresponding to $\tau_{\rm loss} \sim \tau_{\rm diff}$. This indeed matches observation again! (both directly and indirectly, i.e. from measurements of the secondary radiation produced by cosmic ray $e^{\pm}$).

- For secondary electrons and positrons, produced e.g. by the decay of charged pions produced by cosmic-ray proton collisions with protons in the interstellar medium, the source term corresponds to the $Q_p \sim E^{-2.7}$ spectrum found above. The $e^{\pm}$ spectrum after diffusion and energy losses will then follow the same fate as that of primary particles discussed above: a broken power-law, with a hard low-energy spectrum $\psi_{\text{secondary, low-energy}} \sim E^{-3.4}$ and a softer high-energy tail due to energy losses, $\psi_{\text{secondary, high-energy}} \sim E^{-3.7}$. The key point is that, independently of the value of δ (that, remember, tunes the energy dependence of the diffusion coefficient) and of the primary injection spectrum and of energy, the ratio of secondary to primary species is

$$\frac{\psi_{e^+}}{\psi_{e^-}} \sim E^{-\delta}.$$

The prediction of a declining secondary-to-primary ratio was recently found to be at odds with the observed local positron fraction (see the right panel of Fig. 5.2, from Ref. [119]), a fact that spurred much speculation about the nature of the additional positrons responsible for the upturn in the secondary-to-primary ratio.

There are a couple of special limits in which one can get a simple solution to Eq. (5.2) that are worth remembering because they apply to certain interesting physical situations:

1. **No diffusion**: this case corresponds to physical situations where the energy loss timescale is much shorter than the diffusion timescale: the cosmic rays effectively loose their energy before diffusing. In this case, the asymptotic, steady-state (i.e. after transients have died off) solution to the diffusion Eq. (5.2) is

$$\psi(\vec{x}, E) = \frac{1}{b(E)} \int_E^{E_{\max}} dE' \, Q(\vec{x}, E').$$

Exercise 92. Derive the equation above from Eq. (5.2) in the steady-state, $D_0 \to 0$ limit.

There are numerous circumstances where this is a relevant approximation: for example, when (i) the system is very large, with its physical size much larger than the typical diffusion length associated with the energy-loss time (for example clusters of galaxies), or when (ii) the system is such that the energy loss term is very large (for example, a medium with a very dense radiation fields off of which $e^{\pm}$ can efficiently inverse-Compton scatter, or with large magnetic field inducing large synchrotron radiation energy losses).

Exercise 93. Estimate the diffusion coefficient for which the "no diffusion" approximation is reasonable for a 10 GeV electron, for the case of a dwarf galaxy (linear scale size 1 kpc), a Milky-Way-size galaxy (10 kpc), and a cluster of galaxies (1 Mpc), and compare with=break $D_0 \sim 10^{29}$ cm^2/s.

Exercise 94. Suppose $D_0 \sim 10^{28}(E/\mathrm{GeV})^{0.7}$ cm^2/s. Separately calculate for which values of (i) the average magnetic field and of (ii) the radiation density you could neglect diffusion over a region of typical length scale of 10 pc, for an electron energy of 100 GeV.

2. **Burst-like injection from a point source at time t in the past**: in this case, the relevant (spherically symmetric) solution, neglecting energy losses, is

$$\psi \propto Q \cdot \exp\left(-\left(\frac{r}{r_{\rm diff}}\right)^2\right), \tag{5.4}$$

where r is the source distance, and

$$r_{\rm diff} \simeq \sqrt{D(E) \cdot t}.$$

This second case, a burst-like injection, is especially important in connection with Galactic pulsars as sources of high-energy $e^{\pm}$, a potential explanation to the anomalous rising positron fraction found by Pamela [119] and recently confirmed by the Fermi-LAT [132] and by AMS-02 [120], as pointed out by many (including Yours Truly, who once again is not shy to self-promote his own papers, see e.g. [133]).

What are the requirements on the age and distance of a pulsar that contributes to the Pamela positron anomaly? It is easy to give general arguments: first, the pulsar age must be much shorter than the energy loss time scale for energies as large as about $E_e \sim 100$ GeV, in order to have at least some energetic $e^{\pm}$ around! This implies the condition

$$t_{\rm psr} \ll \tau_{\rm loss} = \frac{E}{b(E)}; \text{ for } E = 100 \text{ GeV},$$

$$\tau_{\rm loss} \sim \frac{100}{10^{-16} \cdot 100^2} \text{ s} \sim 10^{14} \text{ s} \sim 3 \text{ Myr},$$

thus a relatively young pulsar. Now, to avoid the exponential suppression of Eq. (5.4) we must have

$$\sqrt{D(E) \cdot t_{\rm psr}} \gg \text{ distance} \rightarrow \text{distance} \ll (3 \times 10^{28} \cdot 100^{0.7} \cdot 10^{14})^{1/2} \text{ cm}$$
$$\sim 10^{22} \text{ cm} \sim 3 \text{ kpc}.$$

So our candidate pulsar is younger than about a mega-year and closer than a few kilo-parsec. One would also like the pulsar to have enough power injected in electron–positron pairs, but this condition, for such a nearby object, is usually fulfilled. Interestingly, several pulsar candidates exist within the desired age and distance, including possibly a handful of the newly discovered radio-quiet gamma-ray pulsars detected by Fermi-LAT (see e.g. Ref. [134]).

As some of you might be aware of, the dark matter annihilation explanation to the Pamela positron fraction anomaly gathered quite a bit of attention (in fact on the order of 10^3 publications entertain this possibility!). Dark matter as a source of the observed excess high-energy positrons faces various issues, including the following:

- there is no evidence for an associated antiproton excess, thus the dark matter must preferentially pair-annihilate into non-hadronic final states (it must be "leptophilic"),
- diffuse secondary radiation from internal bremsstrahlung and inverse-Compton is not observed,
- the needed pair-annihilation rate,

$$\langle \sigma v \rangle \sim 10^{-24} \frac{\text{cm}^3}{\text{s}} \cdot \left(\frac{m_\chi}{100 \text{ GeV}} \right)^{1.5},$$

  is very large for thermal production, and generically leads to unseen gamma-ray or radio emission,

- a XIII century monk[i] pointed out that *"entia non sunt multiplicanda præter necessitatem"*, and the pulsar explanation works just fine to explain the excess positrons.

Despite these difficulties, theorists from all over the world (including myself! [133]) have proposed models that circumvent some, and sometimes even all difficulties, and show proof that Pamela / AMS-02 might have perhaps detected the first non-gravitational signs of dark matter, providing more and more empirical evidence in favor of the so-called "Redman theorem" [135]:

Any competent theoretician can fit any given theory to any given set of facts.

Is there any robust way of disentangling a dark matter origin from a pulsar origin for the excess positrons? One possibility is to use *directional information*. In the diffusive propagation picture, the average dipolar anisotropy in the direction of a source a distance d away is

$$\Delta(E) \equiv \frac{I_{\max} - I_{\min}}{I_{\max} + I_{\min}} \simeq \frac{\phi_{\text{source}}}{\phi_{\text{background}}} \frac{3D(E)}{c} \frac{2d}{\lambda(E)},$$

where $\lambda(E)$ is a "diffusive area" associated with the positrons propagation from the source (in practice, this is the denominator of the Gaussian kernel for the Green's function of the diffusion equation; for more details, see Refs. [136, 137]). Plugging in numbers for candidate nearby pulsars, one gets at most a 0.1–1% anisotropy. Could this be observable?

In principle, AMS-02 is sensitive to such an anisotropy (see their results [120]). However, there are a couple of complications. First, they might detect a "false positive": suppose the magnetic field structure between us and the pulsar is not random, but rather it "funnels" charged cosmic rays in a given direction. Then we could detect an anisotropy even for a completely isotropic interstellar cosmic-ray population! Second, and to me more worryingly, *heliospheric* effects might blur any anisotropy, as the Reader will estimate in the following exercise.

[i] William of Occam.

> **Exercise 95.** Estimate the Larmor radius as a function of energy for cosmic-ray positrons traveling in a typical heliospheric magnetic field ($B \sim$ few nT) and compare with the size of the Solar system (say, 100 AU). Argue that the Larmor radius is much smaller than the Solar system size for positron energies in the Pamela/AMS-02 excess.

After all, we might be stuck advocating Occam's razor in contenting ourselves with a "pulsar" solution to the positron excess.

5.5 The tiny neutral ones

Detecting neutrinos (from an Italian made-up word that indicates the "tiny neutral one", with an English made-up plural form)[j] is hard. As we mentioned, despite building km^3 size detectors, only two astrophysical neutrino sources have been conclusively[k] observed so far: the Sun, and Supernova 1987A! The flip side of the coin is that astrophysical backgrounds are relatively innocuous (albeit of course cosmic rays produce copious "atmospheric" neutrinos as they hit the atmosphere) if one is to use high-energy neutrinos to search for dark matter. The key idea is that dark matter particles can accrue in celestial bodies until large-enough densities start fueling a steady rate of annihilation, yielding high-energy neutrinos (the other SM products being readily absorbed inside the Sun, leaving no measurable effect; a more exotic possibility is the production of other non-SM particles that could also escape the Sun, and possibly be detectable; for a clever example see e.g. Ref. [139]). Neutrinos are pretty much the only thing produced by dark matter annihilation that can escape the core of a celestial body without losing much energy at all, and get all the way out to our km^3 size detectors. The best bets are the Sun and the Earth, with the former, turns out, much better than (although somewhat complementary to) the latter. Let us now make a few estimates for this process, for the case of dark matter capture and annihilation in the Sun.

The dark matter capture rate in the Sun is, roughly

$$C^{\odot} \sim \phi_\chi \cdot \left(\frac{M_\odot}{m_p}\right) \cdot \sigma_{\chi-p},$$

---

[j]After all if you are fine with "broccoli" and "zucchini", you should also try to use "neutrini"!

[k]For recent limits on neutrino point sources with km^3-size telescopes see Ref. [138].

with the dark matter flux

$$\phi_\chi \sim n_\chi \cdot v_{\rm DM} = \frac{\rho_{\rm DM}}{m_\chi} \cdot v_{\rm DM},$$

with the ratio $M_\odot/m_p$ estimating the number of target nucleons in the Sun, and with the dark matter–nucleon interaction cross section $\sigma_{\chi-p}$ being bound by current experimental limits:

$$\sigma_{\chi-p}^{\text{spin-dependent}} \lesssim 10^{-39}\ \text{cm}^2,$$
$$\sigma_{\chi-p}^{\text{spin-independent}} \lesssim 10^{-44}\ \text{cm}^2.$$

Plugging in the relevant numbers, I find

$$C^\odot \sim \frac{10^{23}}{\text{s}} \left(\frac{\rho_{\rm DM}}{0.3\ \text{GeV/cm}^3}\right) \cdot \left(\frac{v_{\rm DM}}{300\ \text{km/s}}\right) \cdot \left(\frac{100\ \text{GeV}}{m_\chi}\right) \cdot \left(\frac{\sigma_{\chi-p}}{10^{-39}\ \text{cm}^2}\right).$$

Exercise 96. Check the normalization for $C^\odot$ claimed above.

We are interested in the number of dark matter particles in Sun: let us call this number N and write down a differential equation that describes the time evolution of $N(t)$:

$$\frac{dN}{dt} = C^\odot - A^\odot[N(t)]^2 - E^\odot N(t).$$

There are various elements I introduced in the otherwise self-explanatory equation above:

- $E^\odot$ describes the "evaporation" of dark matter particles, something that happens if the particles have a (thermal) velocity comparable with the celestial body's escape velocity. Let us quickly estimate this effect. For the Sun

$$v_{\rm esc}^\odot \simeq 1156\ \frac{\text{km}}{\text{s}} \sim 3 \times 10^{-3}\ c,$$

while the Sun's core temperature (the dark matter particles sink to the center after multiple scattering inside the Sun) is

$$T_{\rm core}^\odot \sim 10^7\ K \sim 1\ \text{keV} \sim m_\chi \cdot v_\chi^2,$$

This gives, for the typical dark matter thermal velocities in the core of the Sun

$$v_\chi \sim c \cdot \left(\frac{1\ \text{keV}}{m_\chi} \right)^{1/2} \gtrsim v_{\text{esc}}^{\odot} \rightarrow m_\chi \lesssim 0.1\ \text{GeV}.$$

Bottom line: for dark matter particles in the "preferred" WIMP mass range we can safely neglect evaporation.

Exercise 97. Estimate evaporation in the case of the Earth: what is the relevant dark matter particle mass range for which evaporation matters in the Earth?

- The annihilation rate

$$A^{\odot} \simeq \frac{\langle \sigma v \rangle}{V_{\text{eff}}},$$

where V_{eff} is an effective volume which depends on where WIMPs live inside the Sun; let us use the following (reasonable) guess for the density profile of the sunk WIMPs in the Sun:

$$n(r) = n_0\ \exp\left(-\frac{m_\chi \phi_{\text{grav}}(r)}{T^{\odot}} \right).$$

Exercise 98. Use statistical mechanics to justify the form of $n(r)$ hypothesized above.

We can choose to estimate V_{eff} by identifying an effective radius R_{eff} corresponding to the condition

$$\frac{m_\chi\ \phi_{\text{grav}}(R_{\text{eff}})}{T^{\odot}} \simeq 1 \quad \rightarrow \quad T^{\odot} \simeq \frac{G_N \rho^{\odot} \frac{4\pi}{3} R_{\text{eff}}^3 m_\chi}{R_{\text{eff}}}$$

$$\rightarrow R_{\text{eff}} \sim 10^9\ \text{cm} \left(\frac{m_\chi}{100\ \text{GeV}} \right)^{1/2}$$

$$\rightarrow V_{\text{eff}} \sim 10^{28}\ \text{cm}^3 \left(\frac{m_\chi}{100\ \text{GeV}} \right)^{3/2}.$$

Remember that the Sun's radius is approximately $R^{\odot} \sim 7 \times 10^{10}$ cm, so this radius is smaller than the Sun's radius for reasonably light WIMPs.

Neglecting evaporation, even I can solve the differential equation above, and calculate the quantity we are really interested in, the annihilation rate in the Sun:

$$\Gamma_A = \frac{1}{2} A^{\odot} [N(t^{\odot})]^2 = \frac{C^{\odot}}{2} \left[\tanh \left(\sqrt{C^{\odot} A^{\odot}}\, t^{\odot} \right) \right]^2,$$

with

$$t^{\odot} \sim 4.5 \text{ Byr} \sim 10^{17} \text{ s},$$

the Sun's age (not to be confused with the Sun's core temperature $T^{\odot}$!). One thing we learn from the solution above is that equilibrium between capture and annihilation is reached if

$$t^{\rm eq} \equiv \frac{1}{\sqrt{C^{\odot} A^{\odot}}} \ll t^{\odot}.$$

Do we expect equilibrium or not, for nominal WIMP parameters? Yes, we do! Let us plug in the numbers and convince ourselves of this fact: first, let us find the required annihilation rate for equilibrium

$$C^{\odot} \sim 10^{23} \text{ s}^{-1} \left(\frac{\sigma_{\chi-p}}{10^{-39} \text{ cm}^2} \right),$$

$$A^{\odot}_{\rm eq} \gg \frac{1}{(t^{\odot})^2 \, C^{\odot}} = \frac{1}{10^{34} \cdot 10^{23} \text{ s}} \sim 10^{-57} \text{ s}^{-1}.$$

Now, for vanilla WIMP dark matter

$$A^{\odot} = 3 \times 10^{-54} \text{ s}^{-1} \left(\frac{\langle \sigma v \rangle}{3 \times 10^{-26} \text{ cm}^3/\text{s}} \right),$$

so equilibrium is reached for $\sigma_{\chi-p}$ as small as about 10^{-41} cm^2.

Exercise 99. Redo this calculation for the case of the Earth and find the critical dark matter–nucleon scattering cross section for equilibrium; note that the relevant scattering cross section in the Earth is spin-independent (as the Earth is mostly made of spin-0 Iron nuclei): do you then expect the equilibrium condition to hold for the flux of neutrinos from the center of the Earth?

If equilibrium is achieved, then

$$\Gamma_A \simeq \frac{C^{\odot}}{2}$$

and we do not care about the pair-annihilation cross section (a unique case in the business of indirect dark matter detection!), while we only care about the cross section for dark matter capture. The resulting flux of neutrinos of flavor f will then be

$$\frac{\mathrm{d}N_{\nu_f}}{\mathrm{d}E_{\nu_f}} = \frac{C^{\odot}}{8\pi(D^{\odot})^2}\left(\frac{\mathrm{d}N_{\nu_f}}{\mathrm{d}E_{\nu_f}}\right)_{\mathrm{inj}},$$

where the last factor with the subscript "inj" is the "injection" spectrum of neutrinos per annihilation. Effects that complicate this discussion include neutrino oscillation, absorption of neutrinos in the Sun, and many others that smart people out there have already kindly worked out for you.

The final step is to count the number of events we expect at IceCube or at any other mega-neutrino-detector (these detectors are fundamentally arrays of photomultipliers reading Cherenkov light from muons produced by ν_μ charged-current interactions):

$$N_{\mathrm{events}} = \int \mathrm{d}E_{\nu_\mu} \int \mathrm{d}y \left(A_{\mathrm{eff}} \cdot \frac{\mathrm{d}N_{\nu_\mu}}{\mathrm{d}E_{\nu_\mu}} \cdot \frac{\mathrm{d}\sigma}{\mathrm{d}y}(E_{\nu_\mu}, y) \cdot \left(R_\mu(E_{\nu_\mu})\right)\right),$$

where y indicates the ν_μ energy fraction transferred to the μ in the charged current interaction ($\mathrm{d}\sigma/\mathrm{d}y$) is the relevant cross section for charged current interactions, and the last factor R_μ indicates the muon range in the relevant material the detector lives in (for example, Antarctic ice for IceCube).

The most promising dark matter pair-annihilation final states in this business are those producing a "hard" spectrum of muon neutrinos, i.e. energetic neutrinos. These by all means include W^+W^- and ZZ pairs, that dump out prompt muon neutrinos from the leptonic decay modes of the gauge bosons; luckily, for example in supersymmetry, these are exactly the preferred final states for wino- and higgsino-like lightest neutralinos. Other promising possibilities include "leptophilic" dark matter (that likes to pair-annihilate into leptons, that is, see e.g. Ref. [140]).

The typical flux sensitivity threshold we want to hit to get an interesting signal is about hundreds of muons per km-squared per year, and the typical energy thresholds are 100 GeV for IceCube, which is improved down to 10 GeV for DeepCore and that could go down to the order of a GeV for the further thickly instrumented portion of the detector to be named PINGU.

5.6 Light from dark matter

There are two key ways to get light out of dark matter:

(i) *Prompt* photons from the annihilation or decay event, and
(ii) *Secondary* photons from radiative processes associated with the stable, charged particles produced by the dark matter annihilation or decay event (in practice, the most important ones are electrons and positrons).

Prompt photons are produced either by the two-photon decay of neutral pions $\pi^0 \to \gamma\gamma$ dumped by the hadronization chain of strongly interacting annihilation products, or by internal bremsstrahlung off of charged particles in the intermediate or final state; this second contribution is typically "harder", i.e. more energetic, than the first one. Gamma rays from neutral pion decay have the nice spectral feature that I ask you to derive in the next exercise.

> **Exercise 100.** Show that, independent of the π^0 spectrum, the differential spectrum of gamma rays resulting from $\pi^0 \to \gamma\gamma$, $\mathrm{d}N_\gamma^{\pi^0}/\mathrm{d}E_\gamma$ is symmetric around $E_\gamma = m_\pi/2$ on a log scale in energy (I have a particular affection to this exercise, as my soon-to-be PhD thesis advisor asked it to me at my PhD oral admission exam).

Secondary photons originate as the counterpart to the key energy loss processes for electrons and positrons we discussed in the previous lecture: inverse-Compton and synchrotron. To qualitatively understand the features of inverse-Compton emission, it is useful to commit to memory the formula for the average energy $\langle E_0' \rangle$ of the up-scattered photon (with an original initial energy E_0) as a function of the Lorentz factor $\gamma_e = E_e/m_e$ of the impinging high-energy electron:

$$\langle E_0' \rangle \sim \frac{4}{3}\gamma_e^2\, E_0.$$

The relevant numbers for E_0 are as follows:

$$\textbf{CMB} : E_0 \sim 2 \times 10^{-4}\ \text{eV},$$
$$\textbf{starlight} : E_0 \sim 1\ \text{eV},$$
$$\textbf{dust} : E_0 \sim 0.01\ \text{eV},$$

where "dust" is dust rescattered starlight; for a typical electron–positron injection energy from WIMP dark matter we thus have

$$E_e \sim \frac{m_\chi}{10} \rightarrow \gamma_e \sim 2 \times 10^4 \left(\frac{m_\chi}{100\ \mathrm{GeV}}\right)$$

and

$$E'_{\mathrm{CMB}} \sim 10^5\ \mathrm{eV} \left(\frac{m_\chi}{100\ \mathrm{GeV}}\right)^2.$$

Inverse-Compton emission from dark matter therefore produces hard X-ray photons in the hundreds of keV range. This is great news, as a brand new NASA telescope, NuSTAR, is looking at the sky exactly in that energy range [141]! The inverse-Compton light from starlight and dust falls, instead, in the low-energy gamma-ray regime.

In the monochromatic approximation, synchrotron emission peaks approximately at [135]

$$\frac{\nu_{\mathrm{sync}}}{\mathrm{MHz}} \simeq 10 \cdot \left(\frac{E_e}{\mathrm{GeV}}\right)^2 \left(\frac{B}{\mu\mathrm{G}}\right) \simeq 2.8 \cdot \left(\frac{\gamma_e}{1000}\right)^2 \left(\frac{B}{\mu\mathrm{G}}\right),$$

and the synchrotron power scales like B^2. Dark matter annihilation thus produces a rich, multi-wavelength emission spectrum that goes well beyond the gamma-ray band. An example of the spectrum expected e.g. from the nearby Coma cluster of galaxies is shown in Fig. 5.3, left, from Ref. [142]. Note that the various secondary emission peaks appear exactly where the formulae above would predict them to be!

While secondary emission is always present, it involves the additional steps of accounting for the diffusion and energy losses of the $e^\pm$ produced by dark matter annihilation. Prompt gamma-ray emission, on the other hand, is simpler, and it only involves identifying a dark matter structure and a particle dark matter model; we will thus here make a few estimates for this prompt emission only, which for nominal WIMPs produces photons in the gamma-ray energy range. Also, for definiteness we will talk about dark matter annihilation — dark matter decay is even simpler!

What are the optimal targets and the expected detection rates for gamma ray searches for dark matter? The flux of photons produced by dark matter annihilation from a given direction ψ in the sky and from within a solid angle $\Delta\Omega$ is

$$\phi_\gamma = \frac{\Delta\Omega}{4\pi} \left\{ \frac{1}{\Delta\Omega} \int \mathrm{d}\Omega \int \mathrm{d}l(\psi)\ (\rho_{\mathrm{DM}})^2 \right\} \frac{\langle\sigma v\rangle}{2m_\chi^2} \frac{\mathrm{d}N_\gamma}{\mathrm{d}E_\gamma},$$

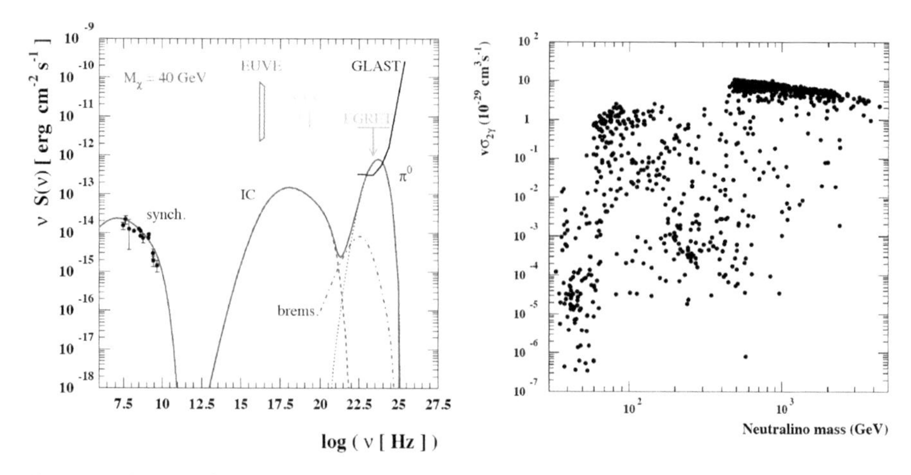

Figure 5.3: Left: The multi-wavelength emission spectrum from the pair-annihilation of a dark matter particle with $m_\chi = 40$ GeV in the Coma cluster of galaxies, from Ref. [142]. Right: the pair-annihilation cross section into two photons for minimal supersymmetric extension of standard model (MSSM) neutralinos, from Ref. [143].

where the last factor is, in fact, a sum of the prompt gamma-ray spectrum from every possible annihilation final state f:

$$\frac{dN_\gamma}{dE_\gamma} = \sum_f \frac{dN_\gamma^f}{dE_\gamma},$$

and where the term in curly brackets is often referred to as the "J factor", a function of the solid angle $\Delta\Omega$ and of course of the direction in the sky, $J = J(\Delta\Omega, \psi)$; it carries units of GeV^2/cm^5. The solid angle $\Delta\Omega$ should be optimized for a given gamma-ray detector and for a given target, field of view and angular resolution to maximize (typically) the signal to noise. It turns out that the relevant solid angles correspond to an angular extent of about one degree, or $\Delta\Omega \sim 10^{-3}$ sr for the Fermi-LAT at an energy of about a GeV, down to an angular extent of 0.1 degrees, or $\Delta\Omega \sim 10^{-5}$ sr for ACT, or for Fermi in the high-energy regime.

Let me give you a "laundry list" of potential interesting targets to search for a gamma-ray signal from dark matter; for most of these targets the "J factor" is approximately the same for a solid angle corresponding to 1 deg or 0.1 deg:

1. Dwarf Spheroidal Galaxies
    - Draco, $J \sim 10^{19}$ GeV^2/cm^5, $\pm$ a factor 1.5;

- Ursa Minor, $J \sim 10^{19}$ GeV^2/cm^5, $\pm$ a factor 1.5;
- Segue, $J \sim 10^{20}$ GeV^2/cm^5, $\pm$ a factor 3.

2. Local Milky-Way-like galaxies
   - M31, $J \sim 10^{20}$ GeV^2/cm^5.
3. Local clusters of galaxies
   - Fornax, $J \sim 10^{18}$ GeV^2/cm^5;
   - Coma, $J \sim 10^{17}$ GeV^2/cm^5;
   - Bullet, $J \sim 10^{14}$ GeV^2/cm^5.
4. Galactic center
   - 0.1°: $J \sim 10^{22} \ldots 10^{25}$ GeV^2/cm^5;
   - 1°: $J \sim 10^{22} \ldots 10^{24}$ GeV^2/cm^5.

To have a detection, we need to have enough photon counts (possibly, and hopefully, a lot!):

$$N_\gamma \sim \int_{E_\gamma \text{ range}} \mathrm{d}E_\gamma \, \phi_\gamma \cdot A_{\text{eff}}(E_\gamma) \cdot T_{\text{obs}}.$$

Table 5.1 gives a rule of thumb for the relevant energy ranges, effective areas and observing time for current and future gamma-ray observatories:

It is instructive to calculate the minimal J factor needed to get at least *some* gamma-ray signal from dark matter. Consider for example Fermi-LAT: over the LAT energy range, typically

$$\int \mathrm{d}E_\gamma \frac{\mathrm{d}N_\gamma}{\mathrm{d}E_\gamma} \sim \frac{m_\chi}{\text{GeV}},$$

so that

$$\phi_\gamma = (\Delta\Omega \cdot J) \frac{1}{8\pi} \frac{\langle \sigma v \rangle}{m_\chi^2} \cdot m_\chi \sim 10^{-32} \frac{1}{\text{cm}^2\text{ s}} \left(\frac{J}{\text{GeV}^2/\text{cm}^5} \right)$$

Table 5.1: Approximate guidelines for the gamma-ray energy range, effective area and typical observation times for the Fermi-LAT, for the ground-based Cherenkov telescope H.E.S.S., and for the future CTA.

	Fermi-LAT	H.E.S.S.	CTA
E_γ range	0.1–300 GeV	0.1–10 TeV	10 GeV–10 TeV
A_{eff}	~ 1 m^2	$\sim 10^5$ m^2	$\gtrsim 10^6$ m^2
T_{obs}	$\sim 10^8$ s	$\sim 10^6$ s	$\sim 10^6$ s

and

$$N_\gamma \sim A_{\rm eff} \cdot T_{\rm obs} \cdot \phi_\gamma \sim 10^{-20} \frac{J}{\rm GeV^2/cm^5},$$

where we put in the nominal values for the effective area and observing time as in the Table 5.1, so, we want

$$J \gtrsim 10^{20}\ {\rm GeV^2/cm^5}.$$

This is a bit bigger than the individual dwarf spheroidal galaxies' J factors quoted above. In fact, combining observations of all (non-detected) dwarf galaxies gives one of the tightest (in my personal opinion *the* tightest, and likely most robust) constraints to date on the dark matter pair-annihilation rate as a function of mass: dwarf galaxies are a virtually background free target, with (stacking a few dwarf galaxies)

$$J^{\rm tot} \sim {\rm few} \times 10^{20}\ {\rm GeV^2/cm^5},$$

therefore the resulting limits [144] are

$$\langle \sigma v \rangle_{\rm lim} \sim 3 \times 10^{-26} \frac{\rm cm^3}{\rm s} \left(\frac{30\ {\rm GeV}}{m_\chi} \right).$$

Interestingly, this rough estimate is in pretty good agreement with the detailed analysis and results of Ref. [144].

A clear-cut signal of dark matter annihilation that could be detected with gamma-ray detectors is the close-to-monochromatic line from the direct annihilation

$$\chi\chi \to \gamma\gamma.$$

Since $E_\chi \sim m_\chi$, $E_\gamma \simeq m_\chi$ and the line is almost monochromatic. Dark matter particles, being dark, ought better not directly couple to photons, and the naive expectation is that the $\gamma\gamma$ amplitude be loop-suppressed, i.e.

$$\frac{\langle \sigma v \rangle_{\gamma\gamma}}{\langle \sigma v \rangle_{\rm tot}} \sim \frac{\alpha^2}{16\pi^2}.$$

This naive estimate gives the correct ballpark over a wide range of parameters in the MSSM, as Fig. 5.3, right (taken from Ref. [143]), illustrates (but things can get a lot worse, and not much better, than the naive estimate!).

A monochromatic line at around 130 GeV was claimed in early Fermi-LAT data [145, 146]. Unfortunately, the line turned out to be a statistical fluke — the significance dropped with time, and the signal did not survive the new generation LAT analysis software, Pass 8. Yet, in my opinion the only conclusive avenue to detect dark matter with gamma rays remains the detection of a *diffuse, monochromatic* (i.e. line) signal [147]. Notice I insist on the morphological aspects: you need a diffuse emission to convince yourself that the line does not originate from a "cold" pulsar wind (very narrow electron energy distribution, giving rise to a nearly monochromatic line from inverse Compton up-scattering of the pulsar's optical emission photons) [148].

Other distinctive spectral features that could arise from dark matter have been discussed and are possible. One is a box-shaped spectrum, for example from the cascade process of dark matter pair-annihilating into a pair of scalars ϕ, in turn decaying into two photons[1] [150]. The following exercise walks you through the resulting gamma-ray spectrum.

Exercise 101. Suppose the dark matter particle χ pair-annihilates to a pair of scalars ϕ,

$$\chi\chi \to \phi\phi,$$

and the scalars decay to two photons, $\phi \to \gamma\gamma$.

(i) Show that the the photons' energy is

$$E_\gamma = \frac{m_\phi^2}{2m_\chi}\left(1 - \cos\theta\sqrt{1 - \frac{m_\phi^2}{m_\chi^2}}\right)^{-1},$$

where θ is the angle between the outgoing photon and the parent scalar in the "lab" frame.

(ii) Show that the resulting spectrum is constant between the energy endpoints

$$E_\pm = \frac{m_\chi}{2}\left(1 \pm \sqrt{1 - \frac{m_\phi^2}{m_\chi^2}}\right),$$

(*Continued*)

[1]Such is the case for dark matter pair-annihilating into neutral pions, or to a pair of axions [149], for example.

Exercise 101. (*Continued*)

and that it has the form

$$\frac{dN_\gamma}{dE_\gamma} = \frac{4}{\Delta E}\Theta(E_\gamma - E_-)\Theta(E_+ - E_\gamma),$$

where $\Delta E = E_+ - E_-$, and Θ is the Heaviside function (for details see Ref. [150]).

Chapter 6

Searching for Dark Matter with Particle Colliders

6.1 Praeludium

The existence of dark matter and the absence of a suitable particle candidate for it within the Standard Model (SM) of particle physics point to the inevitable conclusion that the dark matter belongs to a new "sector" in the realm of particle physics. As such, it makes sense to use colliders — the best tools to explore elementary particles — to search for dark matter.

Can we ever hope to study dark matter particles with colliders in the same way in which, in the past, people have studied, say, new quarks or leptons, or even weakly-interacting neutrinos? Probably not, at least in most dark matter models, as this simple exercise illustrates.

Exercise 102. We want to compare the Galactic dark matter flux at Earth with the flux of dark matter that might be expected from collider production.

Assume $\rho_{\rm DM}(r_\oplus) = 0.3$ GeV/cm^3, $v = 220$ km/s. Assume that an Large Hadron Collider (LHC) detector has an instantaneous luminosity $\dot{\mathcal{L}} = 5 \times 10^{33}$ cm^{-2}s^{-1}.

(i) Get an expression for the flux of dark matter particles produced at the LHC detector under consideration at a distance R, assuming isotropic production (discuss how realistic this assumption is), as a function of the total dark matter pair-production cross section $\sigma_{\rm LHC} = \sigma(pp \to \chi\chi + \text{anything})$.

(Continued)

Exercise 102. (*Continued*)

(ii) Assume $m_\chi = 100$ GeV, and a weak-interaction cross section $\sigma_{\rm LHC} = G_F^2 m_\chi^2$. Compare the flux from LHC production at $R = 10$ m with the Galactic flux.
(iii) For which $\sigma_{\rm LHC}$ are the two fluxes comparable?

The exercise illustrates that, naively, dark matter produced at colliders would have such low fluxes that, given direct detection limits it is impractical/impossible to directly detect it. An additional key difference between direct or indirect dark matter searches and collider searches is that whatever particle is produced at colliders, it will be in general unknown whether such particle has anything to do with the Galactic/cosmological dark matter particle. It will even be unclear whether the particle is, at all, long-lived enough to be the dark matter!

The above caveats notwithstanding, much activity and effort have gone into collider searches for dark matter. The key direction has been to search for dark matter by looking at events with significant (transverse)[a] missing energy/momentum, which would be associated with the energy/momentum carried away by the dark matter particle(s). What else should be detected in conjunction with the, typically, pair-produced dark matter particles is model-dependent, although general guidelines can be obtained. In what follows we describe a few popular approaches and their limitations.

Schematically, there are two distinct possibilities as to how to make predictions to search for dark matter at colliders:

(i) a *top-down* approach, where a "complete" theory is given, and distinct, specific search strategies and signals are considered at colliders, and
(ii) a *bottom-up* approach, where interactions of the dark matter with SM particles and fields are approximated either with contact interactions (in the context, that is, of an *effective theory*, see Appendix. A.3), or through more or less economic *simplified models*.

I start this chapter with examples of the top-down approach, specifically supersymmetry and universal extra dimensions (UED) (Sec. 6.2);

[a]In hadron colliders, the momentum of the center-of-mass frame of the underlying event along the beam direction is unknown.

I then discuss examples from the bottom-down approach: purpose and limitations of effective theories (Sec. 6.3), the merit and scope of simplified models (Sec. 6.4), and invisible Higgs decays to dark matter (Sec. 6.5). As with every chapter in this book, this is an incomplete set of topics given the chapter's title, but one that, I hope, will serve the purpose of capturing key physics results and the most important directions in the field.

6.2 Examples from supersymmetry and UED

Supersymmetry has numerous motivations independent from dark matter that make it a compelling case study for the search for new physics with colliders (see Appendix. A.4 for a quick primer focused on one supersymmetric candidate for dark matter, the lightest neutralino). Searches for supersymmetric particles started when Yours Truly was not even in elementary school, with the *Conseil Européen pour la recherche Nucléaire* (CERN), or European Organization for Nuclear Research $Sp\bar{p}S$ collider. The first significant dent in the parameter space considered highly plausible for weak-scale supersymmetry came however with the Large Electron-Positron Collider (LEP) experiments.[b] Data at the "Z pole" (i.e. at a center-of-mass energy $\sqrt{s} \sim m_Z \simeq 91$ GeV) showed that the decay width of the Z to non-SM particles ($\Delta\Gamma_Z$) and the non-SM-like invisible decay width of the Z, $\Delta\Gamma_{\rm inv}$ (i.e. subtracting off the $Z \to \bar{\nu}_f + \nu_f$ modes) prevented many supersymmetric particles to have a mass less than half the mass of the Z.

Exercise 103. Estimate the decay width of the Z to supersymmetric sneutrinos as a function of the sneutrino mass, and compare with the limits $\Delta\Gamma_{\rm inv} < 2$ MeV.

At LEP2 the center-of-mass energy of e^+e^- was increased to $\sqrt{s} \simeq$ 203–208 GeV, essentially ruling out charged supersymmetric particles with masses up to close to half the center-of-mass energy of the collider. The constraints came e.g. from the non-observation of decays $\tilde{f} \to f + \tilde{\chi}_1^0$, where $\tilde{f}$ is the supersymmetric "sfermion" partner of the SM fermion f, and $\tilde{\chi}_1^0$ is the lightest neutralino, which counts as "missing energy/momentum" as it escapes the detector without leaving any signal.

[b]As a result, supersymmetry was declared defunct after LEP operations, only to be resuscitated and then declared defunct again after the first run of the LHC.

Exercise 104. Calculate the cross section for the process $e^+e^- \to \tilde{e}^+\tilde{e}^-$ and compare with the total luminosity of LEP2 for beam energy between 100 and 104.5 GeV of 233 pb^{-1}. If you want you can even be lazy and only consider the photon exchange diagram, but do try to estimate electroweak corrections.

The Tevatron $p\bar{p}$ collider, operating at center-of-mass energies around 2 TeV, probed pair-production of strongly interacting supersymmetric particles such as squarks and gluinos, for example looking for events with a variety of jets and missing transverse energy (in a hadronic collider the center-of-mass energy of the *underlying* parton-level event is not known, so one can only use missing energy/momentum transverse to the beam: the vector sum of all transverse momenta is approximately equal to zero). An issue with these searches is that typically squarks and gluinos decay to the lightest supersymmetric particle via possibly long decay chains, so that the final momentum of the particle that escapes the detector is "degraded". On the up side, though, if electroweak-inos lighter than the strongly interacting superparticles exist, additional hard (i.e. energetic) leptons are produced and can be used to cut on SM background. A key issue is that e.g. $W \to l\nu_l$ gives a substantial hard-lepton background, suppressed, however, for increasing numbers of hard leptons in the event. Requiring e.g. $n \geq 3$ hard leptons greatly suppresses the leptonic W decay background, but it also has small (signal) production rates.

Finally, third generation squarks abundantly produce b quarks in their decay chains, and it is relatively easy to "tag" a b quark: as a result one could search for missing transverse energy, a pair of b quarks, a certain number of jets and a hard lepton.

Exercise 105. Sketch a decay chain of a sbottom and of a stop producing a final state like $b\bar{b}l\nu_l q\bar{q}'$.

Since 2010, the center stage for searches for new physics with colliders has been taken by the LHC. Figure 6.1 gives a bucket list of the many, many search strategies adopted (for example) by the A Large Toroidal LHC Apparatus (ATLAS) Collaboration (one of the two general-purpose experiments, the other one being Compact Muon Solenoid (CMS)) in searching for *supersymmetry* with LHC data. What is the rationale for these searches? As in

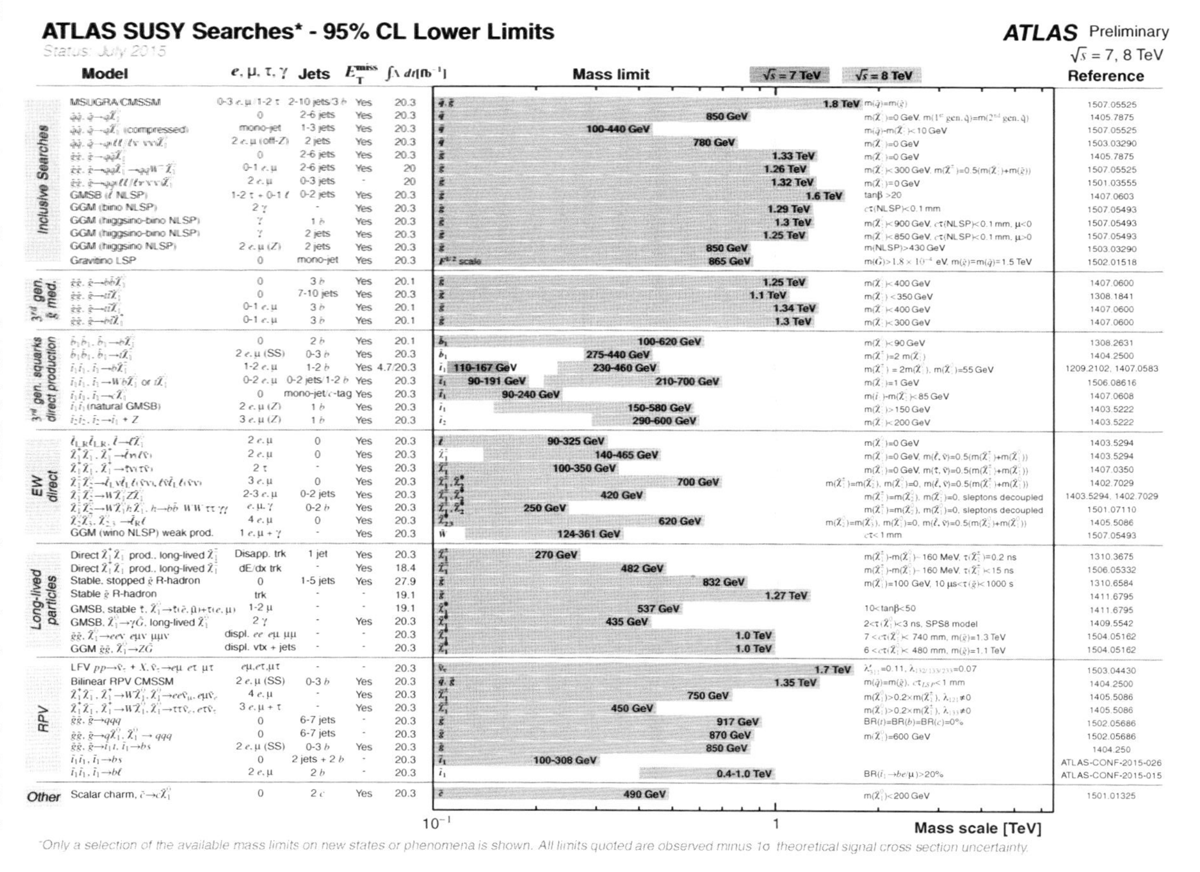

Figure 6.1: A compilation of ATLAS SUSY search results from July 2015, from https://atlas.web.cern.ch, ATLAS Experiment © 2016 CERN.

any experimental search, the key is to maximize the "signal" over "background". In practice, the signal is given by the production rate for the relevant supersymmetric particles, curtailed by whatever cuts are employed to minimize the background — typically mostly given by SM processes (although there also exist SUSY backgrounds to SUSY searches).

As the figure illustrates, searches are based on a variable number of leptons, of a given flavor (e, μ, and/or τ), hard photons, a variable number of jets, and whether or not they triggered on missing transverse momentum. The last column before the exclusion limits gives the total integrated luminosity used in the search. Given a model (the first column), one can calculate a total number of events from the relevant production cross section times the integrated luminosity; each event is then simulated, and the resulting particle yield processed through the cuts ("does the event contain enough missing transverse momentum? leptons etc?"). Simulation of the SM background, fed through the same cuts, gives then a signal-to-noise which is used to set a limit on the mass of the relevant supersymmetric particle. A simple (in principle) algorithm, where, as always, the devil is in the details: what are the cuts that maximize signal-to-background while leave enough "steam" to actually detect *some* signal? etc.

What if a supersymmetric-like excess were ever to be discovered? What would/could we learn about supersymmetry and about supersymmetric dark matter in the short term? A first question is: how could we estimate the mass scale of supersymmetric particles (at a hadron collider, at a next generation e^+e^- collider this question would be void, as we would know the center-of-mass energy exactly...)? A useful quantity that people have studied in quite some detail is "effective mass" $M_{\rm eff}$ given by the magnitude of the momenta of both jets *and* missing energy. If one then plots the distribution in $M_{\rm eff}$ as a function of energy for both background and signal events, the peak of the distribution is found to correlate quite well with the supersymmetry mass scale $M_{\rm SUSY} = \min(m_{\tilde{g}}, m_{\tilde{q}})$, and gives a first handle on the scale of the supersymmetric particle spectra.

The determination of the *individual* mass of supersymmetric particles is trickier, and typically hinges on the fact that there exist kinematic endpoints to the distribution of momenta of particles in an event. For example, consider the *invariant mass* of particle pairs.[c] Think for example of a next-to-lightest neutralino $\tilde{\chi}_2^0$ cascade-decaying to the lightest neutralino plus

[c]Which is just $m_{12} = \sqrt{p_1^2 + p_2^2}$, where p_i are the four-vectors of particles 1 and 2, and the square is of course in the Lorentz sense.

two leptons

$$\tilde{\chi}_2^0 \to l\bar{l}\tilde{\chi}_1^0.$$

The distribution of m_{12} has a maximal value which you are asked to calculate in the next exercise:

Exercise 106.

(i) Show that the dilepton invariant mass is bounded from above by

$$m_{12} \leq m_{\tilde{\chi}_2^0} - m_{\tilde{\chi}_1^0},$$

whether or not the $\tilde{\chi}_2^0$ is produced directly or in cascade decays.

(ii) Illustrate that the end point corresponds to a kinematic configuration where the lightest neutralino is stationary and the two leptons recoil against one another.

(iii) Suppose now that the process involves an intermediate decay step like

$$\tilde{\chi}_2^0 \to \tilde{l}\,\bar{l} \to l\,\tilde{\chi}_1^0\bar{l}.$$

Argue that in this case the lightest neutralino cannot be stationary, and that the "kinematic edge" then shifts to

$$m_{12} < m_{\tilde{\chi}_1^0}\sqrt{1 - \frac{m_{\tilde{l}}}{m_{\tilde{\chi}_2^0}}}\sqrt{1 - \frac{m_{\tilde{\chi}_1^0}}{m_{\tilde{l}}}} \leq m_{\tilde{\chi}_2^0} - m_{\tilde{\chi}_1^0}.$$

Invariant masses can be of course constructed out of an arbitrary number of particles. Consider for example the decay chain

$$\tilde{q}_L \to q\tilde{\chi}_2^0 \to q\tilde{l}^{\pm}l^{\mp} \to ql^{\pm}l^{\mp}\tilde{\chi}_1^0. \tag{6.1}$$

In this case, it would make a lot of sense to study the invariant mass $m(l\bar{l}q)$ of the $l + \bar{l} + q$ system.

Exercise 107. Show that for the decay in Eq. (6.1)

$$m(l\bar{l}q) \leq m_{\tilde{q}}\sqrt{1 - \frac{m_{\tilde{\chi}_2^0}}{m_{\tilde{q}}}}\sqrt{1 - \frac{m_{\tilde{\chi}_1^0}}{m_{\tilde{\chi}_2^0}}}.$$

As an illustrative case of an alternate theory with a complete spectrum (although technically *not* UV-complete), let us consider UED (see Appendix. A.5). In its minimal five-dimensional incarnation, the particle spectrum is fixed by the mass of SM particles and by calculable radiative corrections and boundary terms. The production cross sections at hadronic colliders are then dominated by the pair production of strongly-interacting particles (Kaluza–Klein quarks and gluons), which cascade-decay to the lightest Kaluza–Klein particle (LKP) in a calculable way for a given LKP mass (see Fig. 6.2). The production cross section scales as

$$\sigma(pp \to 2\text{UED particles} + \text{anything}) \sim \frac{\alpha_s^2}{1/R^2},$$

where $1/R$, the inverse compactification radius, sets the mass scale of the spectrum.

It turns out, for example, that at the Tevatron the best limits were obtained searching for 3 or more charged leptons, plus missing transverse energy, resulting in limits just a bit shy of $1/R < 300\,\text{GeV}$. At the LHC, searches were performed both for multi-leptons and for no leptons, resulting in the constraints shown in Fig. A.1, left, for a center-of-mass energy of 8 TeV and an integrated luminosity of around 20 fb^{-1}, [152]. The future sensitivity of LHC searches is shown in the right panel: the best sensitivity

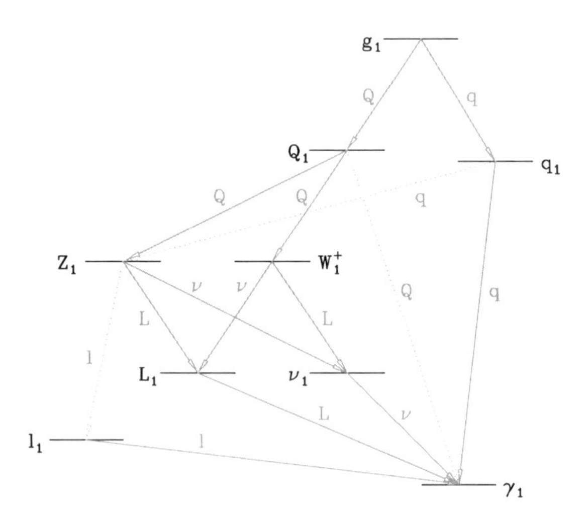

Figure 6.2: The $n = 1$ KK decay chain, showing the dominant (solid) and subdominant (dotted) transitions and the resulting decay products.
Source: Adapted from Ref. [151].

is projected to correspond to searches for events producing four leptons, in conjunction with jets and missing energy [153].

6.3 Effective theory approach

What if the dark matter is effectively the only light "new physics" state accessible at colliders? Then the strategies outlined in the Sec. 6.2, based on the complete knowledge of particle masses and interactions of a given theory, fail: No decay chains or kinematic edges are available, etc. In this case, a first possible approach is to assume that the particles mediating interactions between the dark matter particle χ and the SM are heavy and can be "integrated out", producing an *effective theory* of dark matter interactions with SM particles (see Appendix A.3). In practice, we deal with the dark matter particle (whose mass and spin we get to pick) and with the relevant effective, non-renormalizable operators, each with a certain suppression scale Λ inherited from the mass scale of the heavy mediator that we have integrated out.

A minimal set of assumptions is that the dark matter χ be, for simplicity, a scalar or fermion, odd under some Z_2 symmetry (under which all SM particles are even) which ensures the χ's stability, and that χ is a singlet under all SM gauge interactions. This leaves one.[d] renormalizable "portal" to the SM, the Higgs bilinear $|H|^2$ — we discuss this possibility in Chapter 9, Secs. 9.6 and 6.5. Barring this possibility, the lowest-dimensional operators connecting the dark matter with, for example, a SM fermion bilinear are objects like $\bar{f}\Gamma g$, where f and g are fermions, and Γ is a 4×4 matrix in spinor space that can be built as a linear combination of elements in the following set:

$$\Gamma = \text{span}\{1, \gamma^5, \gamma^\mu, \gamma^\mu\gamma^5, \sigma^{\mu\nu}\}.$$

Terms with fermion derivatives would lead to higher-dimensional operators. Finally, there are operators connecting χ to the massless gauge fields, for example electric and magnetic dipole operators and combinations of $G_{\mu\nu}$ and of its dual gluon field strength tensor.

---

[d]Technically, there is also a "sterile neutrino" N portal through the operator $y_N LHN$, which is, however, lepton number violating, and which allows for dark matter decay, although possibly on long-enough time scales for it to be a viable dark matter candidate; the "kinetic mixing" portal would also exist if the particle were a vector boson, but the resulting dark matter candidate would only be long-lived enough if the vector boson mass were very light, below the electron mass, see Sec. 9.6.

Reference [154] gives a fairly complete list of such operators, for a Dirac fermion χ and for a real or complex singlet. In particular, they *focus* on a comparison between direct detection and collider searches, and choose the coefficients of the effective operator in order to simplify the comparison with direct detection experiments[e]: for example, for the $\bar{q}q$ quark bilinear, the relevant matrix element at low momentum transfer is $\langle N|m_q\bar{q}q|N\rangle$ (where N is a nucleon), thus the normalization factor for the dimension-6 operator $\bar{\chi}\chi\bar{q}q$, where χ is a Dirac fermion, is m_q/Λ^3 rather than $1/\Lambda^2$; similarly, for the quark bilinear $\bar{q}\gamma^\mu q$ the nucleon matrix element is just $\langle N|\bar{q}\gamma^\mu q|N\rangle$, thus the coefficient in front of $\bar{\chi}\gamma_\mu\chi\bar{q}\gamma^\mu q$ is just $1/\Lambda^2$; finally, for the gluon operators, the matrix element is given by $\langle N|\alpha_s GG|N\rangle$, thus a natural choice for the coefficient of the operator $\bar{\chi}\chi G_{\mu\nu}G^{\mu\nu}$ is α_s/Λ^3.

Exercise 108. List all possible lowest-dimensional operators for a Dirac fermion and for a real and complex scalar according to the conventions above for the coefficients (see Table I of [154] for a solution to this exercise).

What is the collider signature associated with the effective operators above? Since we are assuming χ is the only new particle accessible, then the processes of interest are of the type

$$p\bar{p}(\text{or } pp) \rightarrow \bar{\chi}\chi + X,$$

with X whatever SM particles are produced in the reaction; the observable signature will be missing energy from the χ escaping the detector, plus jets or photons (or even Z, W, or h) from the initial state. One of the obvious irreducible, large SM backgrounds is $Z \rightarrow \bar{\nu}\nu$ with associated initial radiation, as well as other processes such as $W \rightarrow l\nu$. The algorithm is otherwise as before: calculate the production cross section, simulate events, devise the best possible set of detector cuts, compare signal and background. As a result, one determines which values of Λ are low enough to be excluded by data, or to potentially be producing a statistically significant signal.

The comparison of collider exclusion limits with direct detection, for a given effective operator, is straightforward.[f] As my collaborators and

[e]Their choices also make it easy to compare effective operators with plausible UV completions.

[f]Francesco D'Eramo would add: "...if you run!"

I pointed out in Ref. [155], the comparison between collider exclusion limits and *indirect* detection searches is, instead, trickier.

Exercise 109. Find which of the following operators are suppressed by chirality (i.e. they give rise to a pair-annihilation amplitude which is proportional to the fermion mass) and/or by p-wave suppression:

$$\begin{aligned}
\mathcal{O}_S &= \bar{\chi}\chi\bar{q}q; \\
\mathcal{O}_P &= \bar{\chi}\gamma^5\chi\bar{q}\gamma^5 q; \\
\mathcal{O}_V &= \bar{\chi}\gamma^\mu\chi\bar{q}\gamma_\mu q; \\
\mathcal{O}_A &= \bar{\chi}\gamma^\mu\gamma^5\chi\bar{q}\gamma_\mu\gamma^5 q; \\
\mathcal{O}_T &= \bar{\chi}\sigma^{\mu\nu}\chi\bar{q}\sigma_{\mu\nu}q.
\end{aligned}$$

See Table I of Ref. [155] for the solution.

A crucial question in the context of effective theories is: what is the range of validity of the effective theory? It is clear that if we are thinking about direct detection experiments, then the relevant energy scale is very low (the momentum transfer, as we saw in Chapter 4, is typically much smaller than the dark matter mass) and the effective field theory description is, likely, fine. At colliders, if the mediator can be produced *on-shell* then clearly the effective field theory description is not a good description: the mediator mass must be much larger than the typical energy of the relevant reaction. For indirect detection, the energy scale depends on the mass of the dark matter particle, and thus the validity of the effective theory on the ratio of the dark matter to the mediator mass.

How do we relate this to the "suppression scale" of a given effective operator? The simplest possibility to map the effective theory to the "UV" is to have a single mediator of mass M and two couplings $g_{1,2}$ to the dark matter particle χ and to SM particles, respectively. Then the effective theory suppression scale

$$\Lambda \sim M/\sqrt{g_1 g_2}.$$

If the effective theory is to describe dark matter pair-annihilation in the early- or late-universe, then $M > 2m_\chi$; on the other hand, if the theory is to be perturbative, then $g_1 g_2 \lesssim (4\pi)^2$. This implies that a weakly-coupled UV completion forces $m_\chi \lesssim 2\pi\Lambda$; on the other hand, the effective theory will break down if the characteristic energy of the reaction, say the momentum transfer $P_{\rm tr}$, is smaller than $4\pi\Lambda$.

In practice, the effective theory truncates the infinite series expansion for the propagator

$$\frac{1}{P_{\rm tr}^2 - M^2} = -\frac{1}{M^2}\left(1 + \frac{P_{\rm tr}^2}{M^2} + \mathcal{O}\left(\frac{P_{\rm tr}^4}{M^4}\right)\right).$$

As such, a good indication of whether or not we are making a large error in making use of the effective theory is to calculate the ratio [156]

$$R(\Lambda) = \frac{\sigma(P_{\rm tr} < 4\pi\Lambda)}{\sigma(\text{any } P_{\rm tr})},$$

of the cross section in the region where the effective theory is legitimate to the total cross section. Taking for example the $\mathcal{O}_S$ operator of Exercise 109, for $m_\chi \lesssim 100$ GeV, Ref. [156] finds that $R(600~\text{GeV}) \simeq 10\%$, $R(1000~\text{GeV}) \simeq 25\%$ and $R(2000~\text{GeV}) \simeq 75\%$, for 14 TeV LHC collisions with cuts similar to those employed to set limits on effective operators. Since LHC limits for that particular effective operator are typically below a TeV, this means that the effective field theory description is largely inappropriate, begging for a description that incorporates, at least schematically, the UV physics of the dark matter-SM mediator.

6.4 Simplified models

Simplified models were devised to mediate between a complete model description of a given DM theory and an effective field theory description, which, as we argued above, is often ill-suited for collider searches. Ref. [157] gives a pretty picture of where simplified models live in "dark matter theory space", which I reproduce in Fig. 6.3

A simplified DM model, besides a dark matter particle which is stable and weakly-interacting enough to escape the detector, contains at least one mediator that couples the DM particle to the SM. Simplified models typically contain most renormalizable terms consistent with Lorentz invariance, SM gauge invariance, and DM stability [157]. Additionally, one might require the simplified model to not violate the exact and approximate accidental global symmetries of the SM — baryon and lepton number, custodial and flavor symmetries should not be strongly broken.

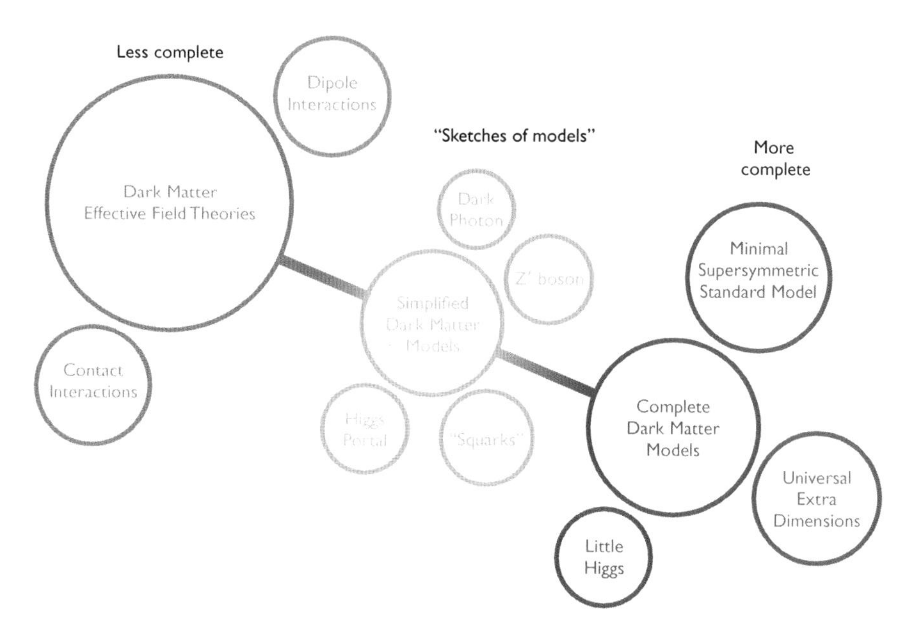

Figure 6.3: Where simplified models live in dark matter theory space. *Source*: Adapted from Ref. [157], originally created by Tim Tait.

A variety of possible representative simplified models has been constructed in the literature. For example, for a fermionic DM particle χ, the effective operators $(\bar{\chi}\Gamma^m\chi)(\bar{q}\Gamma_m q)$ can be generated by a simplified model with a scalar (or pseudoscalar) mediator, or a vector (or axialvector) mediator, e.g. with Lagrangian (for the scalar and vector case).

$$\mathcal{L}_S \supset -\frac{1}{2}M_{\text{med}}^2 S^2 - y_\chi S\bar{\chi}\chi - y_q^{ij} S\bar{q}_i q_j \; + \; \text{h.c.},$$

$$\mathcal{L}_V \supset -\frac{1}{2}M_{\text{med}}^2 V_\mu V^\mu - g_\chi V_\mu \bar{\chi}\gamma^\mu \chi - g_q^{ij} V_\mu \bar{q}_i \gamma^\mu q_j \; + \; \text{h.c..}$$

Naturally, flavor constraints, for example, impose tight requirements on the structure of the "Yukawa" matrix y_q^{ij} etc.

Exercise 110. For the simplified model with $\mathcal{L}_S$, calculate:

(i) The thermally averaged cross section times relative velocity

$$\langle\sigma v\rangle(\bar{\chi}\chi \to S \to \bar{f}f)(T);$$

(Continued)

Exercise 110. (*Continued*)

(ii) The direct detection cross section from S exchange (accounting for scattering off of gluons via heavy quark loops, see Chapter 4);
(iii) The mediator decay width, for $S \to \bar{f}f$ and for $S \to \bar{\chi}\chi$ if kinematically allowed.

We will devote some detailed discussion in Chapter 9, Sec. 9.6 to another simplified model, sometimes dubbed "new minimal SM" [158]: a real scalar singlet S coupled to the SM via the Higgs portal, $S^2|H|^2$. A slightly less simple version is one where one has both a real scalar mediator and a fermion dark matter singlet χ; one step up is a fermion singlet and a pair of fermion doublets with opposite hypercharge (singlet–doublet DM) [157, 159, 160]. The list could go on for a while.

Another interesting possibility is models where the mediator is a *colored* scalar, or vector, particle ϕ (the scalar case is something like a squark in supersymmetry). Then, for a fermionic dark matter χ one has a coupling of the type $\chi\phi q$, and either χ or ϕ carries a flavor index (the SUSY-like case has the ϕ, which is then basically a $\tilde{q}$). Again, it is an interesting exercise to calculate pair-annihilation cross sections, direct detection cross sections and production cross section at colliders (the interested Reader will consult Ref. [156] for details).

6.5 Invisible Higgs decays to dark matter

An indirect but powerful strategy to look for Higgs-portal DM is to study the invisible decay width of the Higgs. Suppose the dark matter is a gauge singlet χ of spin 0 ("S"), 1/2 ("f") or 1 ("V") particle coupled to the Higgs via an operator of the schematic form

$$\lambda_{H\chi\chi}\,\bar{\chi}\chi|H|^2,$$

(in the fermion case $\lambda_{H\chi\chi} \to \lambda_{Hff}/\Lambda$), and suppose that the mass of the χ is less than half the Higgs mass, so that the decay $H \to \chi\chi$ is kinematically allowed. The additional decay mode to $\bar{\chi}\chi$ will result in an extra contribution to the Higgs "invisible" decay width,

$$\Gamma_H^{\rm inv} = \frac{{\rm BR}(H \to {\rm invisible})}{1 - {\rm BR}(H \to {\rm invisible})} \times \Gamma_H,$$

where Γ_H is the SM Higgs decay width (i.e. the total decay width *into SM particles*).

Exercise 111. Show that:

$$\Gamma^{\rm inv}_{H\to SS} = \frac{\lambda^2_{HSS} v^2 \beta_S}{64\pi\, m_H},$$

$$\Gamma^{\rm inv}_{H\to VV} = \frac{\lambda^2_{HVV} v^2 m_H^3 \beta_V}{256\pi\, m_V^4}\left(1 - 4\frac{m_V^2}{m_H^2} + 12\frac{m_V^4}{m_H^4}\right),$$

$$\Gamma^{\rm inv}_{H\to ff} = \frac{\lambda^2_{Hff} v^2 m_H \beta_f^3}{32\pi\, \Lambda^2},$$

where $\beta_\chi = \sqrt{1 - 4m_\chi^2/m + H^2}$ and $v/\sqrt{2} = 174$ GeV is the Higgs vacuum expectation value.

In the Higgs-portal model, Higgs exchange is the only channel that mediates a direct detection (scalar) interaction.

Exercise 112. Show that the spin-independent scattering cross sections of Higgs-portal dark matter χ off of a nucleon N are:

$$\sigma^{\rm SI}_{SN} = \frac{\lambda^2_{HSS}}{16\pi m_H^4}\frac{m_N^4 f_N^2}{(m_S + m_N)^2},$$

$$\sigma^{\rm SI}_{VN} = \frac{\lambda^2_{HVV}}{16\pi m_H^4}\frac{m_N^4 f_N^2}{(m_S + m_N)^2},$$

$$\sigma^{\rm SI}_{fN} = \frac{\lambda^2_{Hff}}{16\pi \Lambda^2 m_H^4}\frac{m_N^4 m_f^2 f_N^2}{(m_f + m_N)^2},$$

where f_N is the usual relevant form factor $f_N \simeq 0.33^{+0.30}_{-0.07}$.

Limits, at a given confidence level, on the branching ratio $H \to$ invisible can then be translated into upper bounds on the direct detection scattering cross section. This was kindly done by the ATLAS collaboration [161], using $BR(H \to \text{invisible}) < 0.28$ at 90% C.L., in Fig. 6.4, and compared with a few direct detection experimental results. The shape of the curves has to do with the mass dependence in both the Higgs decay width and the direct detection scattering cross section.

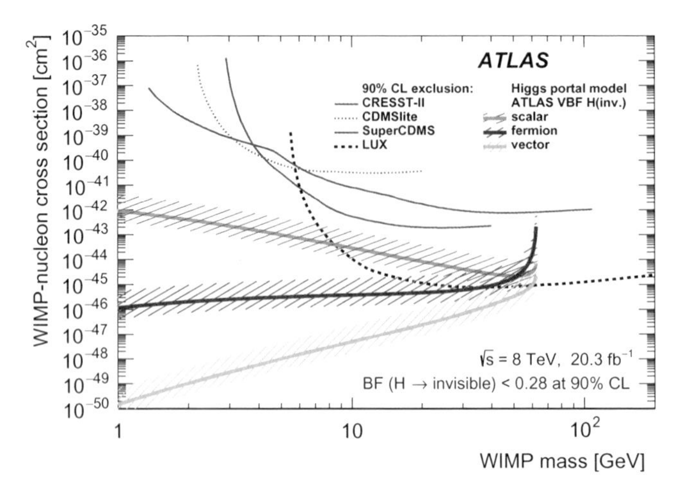

Figure 6.4: Limits on the direct detection spin-independent scattering cross section from the invisible decay width of the Higgs, in a Higgs portal scenario, for three different dark matter spins, from Ref. [161].

Exercise 113. Convince yourself that the dependence on m_χ in Fig. 6.4 makes sense!

Chapter 7

Axions and Axion-like Particles as Dark Matter

7.1 Praeludium

Quantum chromodynamics (QCD), the theory of strong nuclear interactions, has a problem: the *strong CP* problem [162]. The QCD Lagrangian, which explains, as far as we can tell, everything there is to explain concerning strong interactions, reads

$$\mathcal{L}_{\rm QCD} = -\frac{1}{4} G^a_{\mu\nu} G^{a\mu\nu} + \sum_{j=1}^{n} [\bar{q}_j \gamma^\mu i D_\mu q_j - (m_j q^\dagger_{{\rm L}j} q_{{\rm R}j} + {\rm h.c.})] + \frac{\theta g_s^2}{32\pi^2} G^a_{\mu\nu} \tilde{G}^{a\mu\nu}, \quad (7.1)$$

with an implicit summation over suppressed color indices and an explicit one over n quark flavors, and where the *dual* of the (gluon) field strength tensor

$$\tilde{G}^{\mu\nu} \equiv \frac{1}{2} \epsilon^{\mu\nu\rho\sigma} G_{\rho\sigma}.$$

A moment's reflection, and memories of your E&M classes, will convince you that the last piece in Eq. (7.1) is a total derivative, and that it therefore does not affect the equations of motions, or other *perturbative* aspects of QCD. A second moment's reflection will reveal that the last term in Eq. (7.1), known as *theta term*, additionally violates the discrete symmetries CP, T, and P.

> **Exercise 114.** Convince yourself that (i) the theta term in Eq. (7.1) is indeed a total derivative (write it as a 4-divergence), and that (ii) it violates CP.

The Adler–Bell–Jackiw anomaly [163, 164] ensures that (i) the θ term must be present, if none of the quark masses vanishes, and that (ii) QCD depends on θ through the combination of parameters

$$\bar{\theta} = \theta + \arg(\det(\mathcal{M})),$$

with $\mathcal{M}$ the quark mass matrix. Plugging in numbers, one can calculate (very) observable consequences of a non-vanishing $\bar{\theta}$. The most stringent probe of "strong" CP violation is the electric dipole moment of the neutron. With chiral perturbation theory techniques (see Appendix A.3) one can calculate that, for a non-vanishing $\bar{\theta}$, the induced neutron electric dipole moment is

$$d_n \simeq 5 \times 10^{-16} \bar{\theta}\, e \text{ cm}. \tag{7.2}$$

A generic/natural $\bar{\theta} \sim \mathcal{O}(1)$ badly violates current experimental constraints,

$$d_n < \text{few} \times 10^{-26}\, e \text{ cm},$$

by about ten orders of magnitude! Why is $\bar{\theta}$ so small, then?

> **Exercise 115.** I have always been mesmerized by the quality of constraints on electric dipole moments. If you simple-mindedly think of them as constraints on the relative center-of-mass to center-of-charge position offset, you get a taste of the level of precision these experiments achieve. Suppose the neutron was the size of the Earth: what is the relative offset corresponding to 10^{-26} cm?

Equation (7.2) makes it clear that if even only one quark mass equals zero, one can redefine fields via a chiral rotation and do away with the unwanted theta term of Eq. (7.1); in the absence of a massless quark, however, the smallness of θ is an open question. And a question that, for a variety of theoretical reasons, should be taken seriously.

Perhaps the most compelling solution to the issue of the smallness of $\bar{\theta}$ is the 1977 Peccei–Quinn (PQ) theory, whereby $\bar{\theta}$ is promoted to a *dynamical* variable driven to zero by its own classical potential. This is done by postulating a global $U_{PQ}(1)$ quasi-symmetry, which is a symmetry of the theory at the Lagrangian (i.e. classical) level, but explicitly broken by the non-perturbative effects which produce the theta term, and spontaneously broken at a scale f_a. Being a spontaneously broken global symmetry, there must be a pseudo[a]–Nambu–Goldstone boson, the "axion" a, associated with $U_{PQ}(1)$. As shown in the one-line calculation of Ref. [165], the ground state of the axion potential drives $\bar{\theta} \to 0$ and solves the strong CP problem. The axion field is related to $\bar{\theta}$ by

$$a = (f_a/N)\bar{\theta},$$

with N the color anomaly of the PQ symmetry ($N = 6$ in the original PQ–Weinberg–Wilczek axion) and $v_a = Nf_a$ the PQ vacuum expectation value.

More generally, one can consider Standard Model extensions with new global $U(1)$ symmetries spontaneously broken by some (hidden) Higgs-type mechanism at a symmetry breaking scale v_h much higher than the electroweak scale. The field $a(x)$ in the "phase" of the hidden complex Higgs field H_h about its vacuum expectation value,

$$H_h(x) = \frac{1}{\sqrt{2}}(v_h + h_h(x))e^{ia(x)/v_h},$$

has a flat classical potential, $V(a) = \text{const.}$, and is therefore massless (at least at the classical level), while the field h_h has a large mass $m_h \propto v_h \gg v \simeq 246$ GeV. The generic interactions of the "axion-like particle" (ALP) a with SM particles are suppressed by the inverse of the large symmetry-breaking scale v_h,

$$\mathcal{L}_{ALP} = \frac{1}{2}\partial_\mu a \partial^\mu a - \frac{\alpha_s}{8\pi} C_{ag} \frac{a}{f_a} G^a_{\mu\nu} \tilde{G}^{a\mu\nu} - \frac{\alpha}{8\pi} C_{a\gamma} \frac{a}{f_a} F_{\mu\nu} \tilde{F}^{\mu\nu} + \frac{1}{2} \frac{C_{af}}{f_a} \partial_\mu a \bar{f} \gamma^\mu \gamma^5 f,$$

with the decay constant $f_a \simeq v_h$. The couplings of the ALP to SM fields are model-dependent. While for the QCD axion the chiral anomaly dictates that $C_{ag} \neq 0$, that might not be the case in general [166, 167]. String theory predicts a plethora of ALPs [168–170], and much of the following discussion generalizes from the QCD axion to generic ALPs.

[a] "Pseudo" because there is a small mass term in the potential due to the mentioned non-perturbative QCD effects.

As mentioned above, the color anomaly of the PQ current produces a mass for the (QCD) axion, of order

$$m_a \sim \frac{\Lambda^2_{\rm QCD}}{f_a} \sim 0.6\ {\rm eV}\left(\frac{10^7\ {\rm GeV}}{f_a}\right);$$

therefore, for f_a between 100 GeV and the Planck scale, the axion mass ranges between

$$10^{-12}\ {\rm eV} \lesssim m_a \lesssim 1\ {\rm MeV}.$$

Current algebra methods give the interaction terms between (QCD) axions and SM fermions in the form

$$\mathcal{L}_{a\bar{f}f} = ig_f \frac{m_f}{(f_a/N)} a\bar{f}\gamma_5 f,$$

as well as, most notably for axion searches, the coupling of axions to two photons (note that normalization conventions differ; here I use those of Ref. [171]),

$$\mathcal{L}_{a\gamma\gamma} = -g_\gamma \frac{\alpha}{\pi}\frac{a}{f_a}\vec{E}\cdot\vec{B}.$$

In both equations, g_f a g_γ are model-dependent coefficients of order 1. The couplings above allow to set a variety of constraints on the (essentially) single parameter of the theory, the axion mass. For example, if the axion is heavier than the $2m_e \simeq 1$ MeV threshold, it would decay into e^+e^- pairs, and would produce unobserved decays like $\pi^+ \to a(e^+e^-)e^+\nu_e$. Below that threshold, axions decay into two photons. Let us estimate the lifetime for that decay mode.

The decay width for $a \to \gamma\gamma$ is

$$d\Gamma = \frac{1}{2m_a}|\mathcal{M}|^2 d\phi,$$

with

$$|\mathcal{M}|^2 \sim \frac{\alpha^2}{\pi^2 f_a^2} m_a^4,$$

forgetting about g_γ^2 (warning — this can be a big factors in certain model realizations!), and

$$d\phi = \frac{d\Omega}{32\pi^2}.$$

All in all, I would estimate the axion decay width to two photons as

$$\tau_{a\to\gamma\gamma} \sim \frac{16\pi^2}{\alpha^2}\frac{\Lambda_{\rm QCD}^4}{m_a^5} \simeq 10^{24}\ {\rm s}\left(\frac{1\ {\rm eV}}{m_a}\right)^5. \tag{7.3}$$

A good dark matter candidate should, at the very minimum, have an abundance in reasonable agreement with the observed dark matter universal density, be long-lived enough to be still around today, and have hidden from searches for dark matter. We leave the problem of axion dark matter production to Sec. 7.3 and axion searches to Sec. 7.4 (Sec. 7.2 is devoted to what axions and ALPs would do to stars, which in some sense is also a possible detection channel), but Eq. (7.3) allows us to estimate right away how light the axion must be to qualify as a potential candidate for dark matter:

$$\tau_U \sim 10^{10} \times (\pi 10^7)\ {\rm s} \lesssim 10^{24}\ {\rm s}\left(\frac{1\ {\rm eV}}{m_a}\right)^5$$

$$\Rightarrow m_a \lesssim 25\ {\rm eV},\ \ f_a \gtrsim 4\times 10^6\ {\rm GeV}.$$

Axions with such small masses/large PQ scales were dubbed "invisible".[b] Yet, impossible (and invisible) is nothing, as they say.

7.2 Axions and stars

An interesting and instructive story is what particles such as axions (light, very weakly coupled) would do to stars — be it tranquil stars like the Sun or violent ones, such as supernovae. Here is a short account of this story.

Stars are long-lived not because of the (super fast) timescales associated with nuclear reactions but, rather, because they are very inefficient at radiating energy. Ask yourself: how long does it take for a photon produced in the center of the Sun to escape the Sun? (We will calculate that timescale shortly) It is a long, long time. On the contrary, if axions exist, they are produced and radiated out of stellar systems rather efficiently: depending on their mass, axions have very long mean free paths in stars (even in dense neutron stars), so they easily free-stream out of stellar systems, carrying

[b] As opposed to more massive axions, or lower f_a scales: for example axions between 10 keV and the e^+e^- threshold were quickly ruled out by searches for rare meson or quarkonium decays, such as $K^+ \to \pi^+ + a$ or $J/\psi \to a + \gamma$ or $\Upsilon \to a + \gamma$, where the axion would escape the detector before decaying into two photons.

away potentially large fractions of the star's "luminosity". This is the key to constraints on axion (and other light particles) models based on stellar evolution.[c]

Right, so what is the timescale for a photon produced at the center of the Sun to escape the Sun? Very, very roughly the mean-free path of a photon in the Sun is $\lambda_\gamma \sim 1/(\rho_\odot \cdot \sigma_T)$. Now, $\rho_\odot \sim 1.4 \times 10^3$ kg/m^3, $\sigma_T \sim 10^{-24}$ cm^{-2}, so $\lambda_\gamma \sim 2$ cm. But this is a diffusive, "Brownian" process, so we need to estimate the relevant diffusion coefficient. A simple-minded estimate for the diffusion coefficient $D \sim \lambda v_p \sim 6 \times 10^8$ cm^2/s, where $v_p \sim \sqrt{T_\odot/m_p}$ is the average target particles' velocity; finally, the time it takes for the photon to get out of the Sun is

$$t \sim R_\odot^2/D \sim 10^7 \text{ years}.$$

Yes, it takes on the order of 10 million years for a photon to escape the Sun. This is a very long time. It is clear that for such an inefficient system at cooling, axions can indeed potentially act as "super-coolants"!

The best super-coolant a has a mean free path λ_a comparable to the size of the object R: if $\lambda_a \gg R$, the production rate is suppressed, while for $\lambda_a \ll R$ we are in the diffusive regime just as for photons.

Exercise 116. Using an average solar density of 1.5 g/cm^3, and assuming an axion-nucleon cross section $\sigma \sim 1/f_a^2$, calculate λ_a as a function of m_a in units of the solar radius $R_\odot \simeq 7 \times 10^{10}$ cm.

The key processes for axions in stars are the Compton-like reaction $\gamma + e \to a + e$ and the axion-bremsstrahlung reaction $e + Z \to a + e + Z$. Both processes are $\propto f_a^{-2} \propto m_a^2$. For example, the energy-loss rate for the Compton-like process is

$$\frac{dE_{\text{a-Compton}}}{dt} \sim n_e n_\gamma \frac{\langle E_a \sigma |v| \rangle}{\rho},$$

a thermal average involving the axion energy and the production cross section times relative velocity, divided by the mass density, times the electron

[c] The definitive guide on this topic is Ref. [172], by the definitive expert Georg Raffelt.

and photon number densities. For the Sun, this gives

$$\dot{\epsilon}_{a,\odot} \sim 10 \left(\frac{m_a}{1\ \text{eV}}\right)^2 \frac{\text{erg}}{\text{g} \cdot \text{s}}. \tag{7.4}$$

For reference, the competing energy loss mechanism in the Sun is

$$4p \to {}^4\text{He} + 2e^+ + 2\nu_e + \gamma,$$

which amounts to a few ergs/(g · s). Thus, if $m_a \gg 1$ eV the Sun would primarily loose energy via axion emission!

The energy loss process in Eq. (7.4) results in an "axion luminosity"

$$L_a \sim \int_{\text{star}} \dot{\epsilon}_{a,\odot} dM,$$

with dM a stellar mass infinitesimal element. For the Sun one gets

$$L_a \sim 6 \times 10^{-4} \left(\frac{m_a}{1\ \text{eV}}\right)^2 L_\odot. \tag{7.5}$$

We will use Eq. (7.5), when we discuss detecting solar axions in the laboratory.

Exercise 117. Derive the order of magnitude for Eqs. (7.4) and (7.5).

Since the solar photon luminosity is whatever is observed, axion emission would result in enhanced nuclear energy production, and therefore a *larger neutrino flux*. Results from the Sudbury Neutrino Observatory (SNO) and the standard solar model imply that the axion luminosity cannot exceed about 10% of the solar luminosity, which gives a bound on $m_a \lesssim 10$ eV. Better limits are obtained from globular clusters: axions would accelerate the consumption of helium in horizontal-branch stares to a level that, for a certain axion mass, would be at odds with observations. Conservatively, Ref. [128] quotes a limit on $m_a \lesssim 1$ eV (the precise numbers depend on the axion model).

Axions could also potentially greatly affect the duration of supernova neutrino bursts. This is a rather interesting story, and the one event we have data on, SN1987A, provides good constraints. The axion luminosity L_a should be compared, and should not exceed, the thermal neutrino luminosity, $L_\nu \sim 10^{53}$ ergs/s. Now, depending on the axion mass, the axion luminosity from a supernova behaves according to two very different regimes. For low axion masses (thus suppressed couplings, generically $\propto m_a$) axions free-stream out of the supernova. The axion luminosity then scales simply as the axion production rate, which is $\propto m_a^2$ as we have seen above, thus $L_a \sim m_a^2$. The right proportionality factor is not hard to estimate, and gives

$$L_a \sim 10^{59}\text{ergs/s}\left(\frac{m_a}{1\text{ eV}}\right)^2.$$

As a result, $L_a \gg L_\nu$ for $m_a \gg 10^{-3}$ eV. However, if the axion mass (and thus the coupling to nucleons) is too large, a physically distinct regime sets on: this second regime kicks in when axions are effectively "trapped" in the neutron star, i.e. their mean free path is comparable to the size of the neutron star. Let us estimate for which axion mass range axions are effectively trapped.

Neutron star reference densities are on the order of 10^{15} g/cm^3, thus the nucleon number density is around $n \sim 6 \times 10^{38}$ per cm^3 (wow!). Let us assume on dimensional grounds that the relevant axion-nucleon cross section is on the order of $\sigma \sim 1/f_a^2$. This gives

$$\lambda_a = \frac{1}{n \cdot \sigma} \sim 10^2\text{cm}^2\left(\frac{1\text{ eV}}{m_a}\right)^2,$$

which implies that trapping occurs for axion masses larger than $m_a \gtrsim 10^{-2}$ eV. Remarkably, it turns out that this is a pretty good estimate. For larger masses, the larger the axion coupling the longer the axion is trapped. It is not hard, but unwarranted here, to go through the details of how to calculate L_a in this regime. The bottom line is $L_a \sim m_a^{-16/11}$ ($m_a \gg 10^{-2}$ eV). Matching the L_a above and below 10^{-2} eV, we get that axions would drastically affect SN burst duration between $10^{-3} \lesssim m_a/(1\text{ eV}) \lesssim 2$.

Interestingly, Ref. [173] finds that for larger masses, while the SN would not have *cooled* via axion emission, enough axions would have been emitted to actually trigger excess events[d] at Kamiokande II. This excludes the window between $20 \lesssim m_a/\text{eV} \lesssim 20{,}000$ (for even more massive axions the flux from SN1987A would have been again too suppressed). A small gap between the two SN1987A bounds (burst duration and "direct" detection at Kamionkande II) remains, but is effectively ruled out by axions being thermally produced as "hot" dark matter particles and contributing excessive extra radiation-like energy density in the Universe. We discuss these thermal axions in the next section.

---

[d]Via the reaction $a + {}^{16}\text{O} \rightarrow {}^{16}\text{O}^*$ and subsequent 5–10 MeV nuclear de-excitation gamma rays.

7.3 Axion production

There are three key mechanisms potentially associated with the production of axions in the early universe: (i) *thermal* production of *hot* axions, and *non-thermal* production of *cold* axions via (ii) the so-called misalignment mechanism or via (iii) cosmic axion strings decay.

Thermal axions

In the early universe, axions are created and annihilated by processes such as

$$a + i \leftrightarrow 1 + 2, \tag{7.6}$$

of cross section σ_i; as usual, the relevant rate for the reaction above has the form

$$\Gamma = \sum_i n_i \langle \sigma_i v \rangle,$$

where the n_i are the number densities of particle species i, and where I should not remind you of what we said about the symbol v in Chapter 3: same here.

> **Exercise 118.** Explain why we neglected the thermal freeze-out of axion pair-annihilation.

The most important processes involve gluons and quarks, e.g. $a+g \leftrightarrow \bar{q}+q$ or $g+g$, or $a+q(\bar{q}) \leftrightarrow g+q(\bar{q})$. Now, let us estimate the rate Γ for such processes. The relevant term comes from the "offending" theta term we started this whole section with.[c] To fix the notation, let us look at that term once again:

$$\frac{g_s^2}{32\pi^2}\frac{a}{f_a}G\tilde{G} = \frac{\alpha_s}{8\pi}\frac{a}{f_a}G\tilde{G},$$

where the strong interactions "fine structure constant" $\alpha_s \equiv g_s^2/(4\pi)$. If I were to estimate the typical cross section $\sigma_{q,g}$ of interest, I would guess

$$\sigma_{q,g} \sim \frac{\alpha_s^3}{\pi^2 f_a^2}.$$

[c] Note that we are concerned with the axion prior to mixing with the η and π^0 mesons here, since the temperatures we have in mind are much larger than the QCD scale.

Note that while I neglected factors of $\mathcal{O}(1)$, I am keeping the π^2. This is for personal reasons.[f] Now, let us estimate the temperature at which the processes of Eq. (7.6) go out of equilibrium:

$$\frac{\Gamma}{H} \sim 1 \Rightarrow N_c N_f T^3 \sigma_{q,g} \sim \frac{T^2}{M_P}.$$

In the equation above, we are supposed to keep track of the color and quark generation factors (symbolically indicated with $N_c N_f$) somehow. Using $\alpha_s \simeq 0.03$, I find the thermal axion (th.ax.) freeze-out temperature to be of the order

$$T_{\text{th.ax.}} \simeq \text{few} \times 10^{11}\ \text{GeV} \left(\frac{f_a}{10^{12}\ \text{GeV}} \right).$$

A Kosher calculation would give few $\simeq 5$ [174].

Exercise 119. Do the rough estimate and try the Kosher calculation of $T_{\text{th.ax.}}$.

Note that thermal axions freeze out at that temperature only if $T \lesssim v_a$ (remember that $v_a = N f_a$ is the PQ v.e.v., and N is the color anomaly of $U_{\text{PQ}}(1)$). We will see later that non-thermal axion production indicates that to avoid overclosing the universe $f_a \lesssim 10^{12}$ GeV, so indeed some axions which around today might have originated from thermal processes, unless they are wiped out by inflation with a reheating temperature $T_{\text{reheating}} < T_{\text{th.ax.}}$.

What about other axion production/annihilation processes that might occur and thermalize at later, lower temperatures? People have postulated that new colored, massive fermions can produce populations of thermal axions [174], but after the QCD phase transition ($T \lesssim 200$ MeV) pions surely play a role via reactions of the form

$$\pi + \pi \leftrightarrow \pi + a.$$

[f] When I was an undergraduate at Scuola Normale, in Pisa, I was traumatized by an Instructor who said very menacing things about forgetting factors of $\mathcal{O}(\pi^2)$: that is a whole order of magnitude mistake right there!

At $T \sim m_\pi$, the relevant cross section is of order $\sigma_\pi \sim 1/f_a^2$, so at $T \sim m_\pi$ the processes above are in equilibrium for

$$\frac{n_\pi \sigma_\pi}{H} \sim \frac{m_\pi M_P}{f_a^2} \gtrsim 1 \quad \Rightarrow f_a \lesssim 5 \times 10^8 \text{ GeV}.$$

Now, this is sort of interesting because (i) that is the regime of interest for axions to be dark matter candidates (see Eq. (7.3)) but also because (ii) that number is *very* close to the bound from SN1987A we discussed above (the burst duration constraint). Also, the relevant temperatures $T \sim m_\pi$ are so low that this thermal axion population could not probably be wiped out by any significant entropy injection episode (e.g. inflation with very low-temperature reheating).

Thermal axions would constitute *hot* dark matter; as such there is one pro and two cons: the pro is that it is easy to calculate the resulting abundance as a function of the axion mass: bam!

$$\Omega_{\text{th.ax.}} \sim \frac{m_a}{130 \text{ eV}}. \tag{7.7}$$

Now for the cons: (1) we know dark matter cannot be predominantly hot. In fact this constrains directly $m_a < 0.7$ eV [175]; (2) for $\Omega_{\text{th.ax.}} \sim \Omega_{\text{Dark Matter}}$, i.e. for $m_a \sim 13$ eV, the axion lifetime would be shorter than the age of the Universe, remember

$$\tau_a \sim \tau_U \left(\frac{19 \text{ eV}}{m_a} \right)^5.$$

So no, if axions are the dark matter, they were definitely not entirely thermally produced. However, they could have been (and, if they exist they likely were) produced non-thermally.

Cold axions: the misalignment mechanism and axion strings

As $\bar{\theta}$ relaxes to 0, it oscillates. As for (almost) any "honest-to-god" potential near a minimum, this potential will be quadratic in $\bar{\theta}$; the curvature of the potential will be m_a^2, so that

$$V(\bar{\theta}) \simeq m_a^2(T) \left(\frac{f_a}{N} \right)^2 \bar{\theta}^2,$$

giving a Lagrangian

$$\mathcal{L} \simeq \left(\frac{f_a}{N}\right)^2 \left(\frac{\dot{\bar{\theta}}^2}{2} - m_a^2(T)\frac{\bar{\theta}^2}{2}\right).$$

There is a nice way to visualize the axion mass, and to convince yourself of the overall factors in the potential.[g] Notice the implicit temperature dependence of the axion mass,

$$m_a(T) \simeq 0.1 m_a(T=0) \left(\frac{\Lambda_{\rm QCD}}{T}\right)^{3.7}, \quad (T \gg \Lambda_{\rm QCD}/\pi) \tag{7.8}$$

implying that the axion is practically massless at temperatures larger than $\Lambda_{\rm QCD}$. The Lagrange equations of motion give

$$\ddot{\bar{\theta}} + 3H\dot{\bar{\theta}} + m_a^2(T)\bar{\theta} = 0.$$

Exercise 120. Derive the equation of motion above.

At $T \gg \Lambda_{\rm QCD}$ the solution to the equation of motion is trivial, $\bar{\theta} = \bar{\theta}_1 =$ constant. This constant is called *"misalignment angle"*. After $m_a(T) \gtrsim 3H$, you have got the nicest differential equation in physics: the harmonic oscillator. $\bar{\theta}$ starts oscillating with a frequency $\sim m_a(T)$. At that point we can use the whole bag of harmonic-oscillator tricks our physics teachers have drilled into our brains year after year after year. For example, substitute $\dot{\bar{\theta}}$ for its average value ρ_a over an oscillation:

$$\rho_a = \left(\frac{f_a}{N}\right)^2 \left(\frac{\dot{\bar{\theta}}^2}{2} + m_a^2(T)\frac{\bar{\theta}^2}{2}\right),$$

and turn the equation of motion for $\bar{\theta}$ into the following equation of motion for ρ_a:

$$\dot{\rho}_a = \left(\frac{\dot{m}_a}{m_a} - 3H\right)\rho_a.$$

[g]It is somewhat tricky as it involves slightly tipping a wine bottle by an angle of $\mathcal{O}(m_a^2/f_a^2)$. Just try. Remember the bottle should be emptied first.

Exercise 121. Slow down and work out the equation of motion for ρ_a.

The solution to the equation above is simple, if one neglects $\dot{m}_a/m_a$ (i.e. in the *adiabatic* approximation i.e. $\dot{m}_a/m_a, H \ll m_a$; note that this approximation can break down, see e.g. [176] for a recent review)

$$\rho_a = \text{const} \times \frac{m_a(T)}{a^3}.$$

So when the axion mass settles to its asymptotic low-temperature value, the axion energy density redshifts like non-relativistic matter! During the mass-varying epoch, however, this is not true. Rather, the axion *number density* per co-moving volume is constant!.[h] This is great, because it gives us a straightforward way to calculate the relic number of axions today from the misalignment angle: the initial energy density at the onset of oscillations is

$$\rho_a = \frac{m_a^2(T_1)}{2}\bar{\theta}_1^2\left(\frac{f_a}{N}\right)^2,$$

where T_1 is defined as the temperature at which $m_a(T_1) = 3H(T_1)$. As a result, $T_1 \sim m_a^{0.18}$, in the regime when Eq. (7.8) applies.

Exercise 122. Show that $T_1 \sim m_a^{0.18}$.

In a standard cosmological setting (no entropy injection), during oscillation the ratio of the axion number density to entropy density is conserved, and given by

$$\frac{n_a}{s} \sim \frac{m_a(T_1)\bar{\theta}_1^2\,(f_a/N)^2}{T_1^3} \sim \frac{\bar{\theta}_1^2 m_\pi^2 f_\pi^2}{m_a^2 T_1 M_P},$$

where in the last step I used $f_a/N \sim f_\pi m_\pi/m_a$. Bottom line is that the cosmological axion relic density from misalignment $\Omega_{\text{mis}} \sim m_a^{-1.18}\bar{\theta}_1^2$. Carefully

[h]Again, this is true only in the adiabatic approximation $\dot{m}_a/m_a, H \ll m_a$; anharmonic terms can be large for large misalignment angles, as pointed out e.g. in [177].

putting numbers in, Ref. [174] gets

$$\Omega_{\rm mis}h^2 \simeq 0.4 \left(\frac{m_a}{10\ \mu\text{eV}}\right)^{-1.18} \left(\frac{\bar{\theta}_1}{\pi}\right)^2.$$

What should we plug in for $\bar{\theta}_1$? A reasonable guess is to use

$$(\bar{\theta}_1)_{\rm RMS} \equiv \left(\int_{-\pi}^{\pi} \mathrm{d}\bar{\theta}_1 \frac{\bar{\theta}_1^2}{2\pi}\right)^{1/2} = \frac{\pi}{\sqrt{3}},$$

which gives

$$\Omega_{\rm mis,RMS}h^2 \sim 0.13 \left(\frac{m_a}{10\ \mu\text{eV}}\right)^{-1.18},$$

implying the "right" axion mass to have axions from misalignment be the dark matter is around 10 μeV (there are of course caveats, for example if the PQ symmetry is broken during inflation, see e.g. [178]).

Are these axions from misalignment cold or hot dark matter particles? To answer this, we simply need to calculate the typical axion velocity today

$$\frac{v_a}{c} \sim \left(\frac{p_a}{m_a}\right)_{\rm today} \sim \frac{p_1}{m_a}\frac{T_0}{T_1} \sim 10^{-18} \left(\frac{m_a}{10\ \mu\text{eV}}\right)^{-0.82},$$

where I assumed $p_1 \sim 1/H(T_1)$. So cold, very, very cold.

The distribution of θ_1 values in the universe need not be a smooth function; generically, it has a non-trivial topology: in other words it cannot be smoothly deformed to a constant value throughout space.[i] As a result, a network of "axion strings" and domain walls form. Let us discuss strings. The energy stored per unit length in these strings depends on the axion properties via $\mu \sim f_a^2 \log(f_a d)$, where d is the typical distance between strings; the energy density stored in the axion strings scales as $\rho_s \sim \mu/t^2$, so that the ratio of the axion string density to radiation density is $\propto \mu$. Over one Hubble time, the energy density of axion strings is dissipated via the only available process, the radiation of axions. As a result, the change in the

[i]Unless inflation occurred before or during the spontaneous breaking of the PQ symmetry.

number density of axions to entropy density is [13]

$$\Delta\frac{n_a}{s} \sim \frac{\mu/t^2}{\omega T^3}\Delta(Ht),$$

where ω is the average energy per axion radiated. Integrating the equation above gives the relic axion density from string decay:

$$\frac{n_a}{s} \sim \mu\int_{T_1}^{f_a}\frac{dT}{\omega M_P^2}.$$

Note that the lower limit of integration stems from the fact that below T_1 the string network becomes a network of domain walls which quickly decays as the axion mass becomes dynamically important. Crucial to the calculation of the relic axion density is thus $\omega(t)$. Different groups obtain different results, including from lattice simulations. We will simply quote here those results for reference, and using the same notation as for the relic axion density from the misalignment angle, we get:

$$\Omega_{\text{strings+domain walls}}h^2 = (3.5 \pm 1.7)\left(\frac{m_a}{10\ \mu\text{eV}}\right)^{-1.18}, \qquad [179]$$

$$\Omega_{\text{strings+domain walls}}h^2 \sim 0.4\left(\frac{m_a}{10\ \mu\text{eV}}\right)^{-1.18}. \qquad [174]$$

In the first case, the favored axion mass would be around 200 μeV, while in the second case around 30 μeV, if string and domain-wall decays were to provide the bulk of axion dark matter. Once again, had the universe undergone reheating at a temperature lower than the PQ breaking scale, then cold axions would presumably be dominantly produced by the misalignment mechanism, with a peculiar value of θ_1 associated with the uniform random "initial" misalignment angle.

7.4 Axion detection

Often mentioned as *the* alternative to weakly interacting massive particles (WIMPs), axions have received well-deserved experimental attention. If one is a theorist, especially a theorist specializing primarily in WIMPs, one should expect the inevitable *No Wimps!* question: "...and what if instead the dark matter is axions?" at any point in one's career.[j] I will

---

[j]For example during job interviews, which is what happened to me. More than once.

discuss four ways to search for ALPs: (1) looking for the decay of axions in astrophysical objects; (2) producing and detecting axions in the laboratory ("shine-through-the-wall" experiments); (3) detecting Galactic axions with microwave cavities; and (4) looking for axions produced in the Sun.

Cosmological axion decay

The flux of (quasi-)monochromatic photons from axions decaying in an object of mass M at a distance d is

$$\phi_\gamma = 2\frac{M}{m_a}\frac{1}{4\pi d^2}\frac{1}{\tau_a}.$$

The (two) photons per axion decay have $E_\gamma \sim m_a/2$, with a (very small) Doppler-broadened line width.

Exercise 123. Estimate the line width for the photons produced by axion decay.

Now, consider a typical cluster of galaxies, $M \sim 10^{15} M_\odot$ and 100 Mpc away, use the expression for the axion decay width into two photons as a function of the axion mass (Eq. (7.3)) and you get

$$\phi_\gamma \sim \frac{10^3}{\mathrm{cm}^2\mathrm{s}}\left(\frac{M}{10^{15}M_\odot}\right)\left(\frac{100\ \mathrm{Mpc}}{d}\right)^2\left(\frac{m_a}{1\ \mathrm{eV}}\right)^4.$$

Exercise 124. Cross check that my estimate of the 10^3 photons per cm^2 per s above is correct.

For cosmologically interesting axions, we are interested in frequencies $\nu \sim \mu$eV, which is around 2.4 GHz: radio frequencies; for thermal ALPs,[k] instead, we are interested in frequencies $\nu \sim$ eV, i.e. wavelengths of order 25,000Å.

Every time we deal with a given frequency, it is good education and good practice to abide by the rules of observers who study those frequencies. Here, we shall need to convert the flux above in units of "specific intensity" I, i.e. ergs $\mathrm{cm}^{-2}\ \mathrm{s}^{-1}\ \text{Å}^{-2}\ \mathrm{arcsec}^{-2}$. This of course requires massaging the

[k]This range of mass is excluded for the QCD axion.

flux above by (1) multiplying by the photon energy squared and converting appropriately energy into wavelength (hint: $\lambda_a \simeq 24,800\text{Å}/(m_a/1\text{ eV})$. Also, we need to divide by the solid angle subtended by the object under consideration (this is roughly $\Delta\Omega \simeq 2\pi\theta^2$, with $\theta \sim 1$ Mpc/100 Mpc). What I find is

$$I \sim 10^{-20}\text{ergs cm}^{-2}\text{s}^{-1}\text{Å}^{-2}\text{arcsec}^{-2}\left(\frac{M}{10^{15}M_\odot}\right)\left(\frac{100\text{ Mpc}}{d}\right)^2\left(\frac{m_a}{1\text{ eV}}\right)^6.$$

A similar flux, but roughly three orders of magnitude smaller, is predicted for axions decaying in the Milky Way halo [180].

Exercise 125. Do the units gymnastics above, and estimate the brightness for the case of the galaxy.

Are these fluxes observable? For optical frequencies, we should compare this line intensity with the brightness of the night sky, roughly

$$I_{\rm NS} \sim 10^{-17}\text{ ergs cm}^{-2}\text{s}^{-1}\text{Å}^{-2}\text{arcsec}^{-2},$$

which implies that ALPs with a mass $m_a \gtrsim 3$ eV should produce a detectable line (again, these guys cannot be QCD axions). This estimate is not different from early estimates [180]; the most recent searches carried out with the Very Large Telescope observations of Abell 2667 and 2390 exclude axions down to a mass of 4.5 eV (unless very suppressed axion-to-two-photon couplings are postulated) [181].

What about radio? The best bet is using nearby dwarf galaxies, which have both very low velocity dispersion (hence producing very narrow axion decay lines) and are very radio-quiet. Observations with the Haystack radio telescope of three dwarfs (LGS 3, Pegasus, and Leo I) have produced constraints in the (narrow) band $298 \lesssim m_a/\mu\text{eV} \lesssim 363$, which bound the effective axion-to-two-photon vertex

$$\mathcal{L}_{a\gamma\gamma} = G_{a\gamma\gamma} a\vec{E}\cdot\vec{B},$$

to $G_{a\gamma\gamma} \lesssim 10^{-9}\text{ GeV}^{-1}$ [182]; unfortunately this limit is not even close to the predicted $G_{a\gamma\gamma} \sim \frac{\alpha}{2\pi}\frac{m_a}{m_\pi f_\pi} \sim 3\times10^{-14}\text{ GeV}^{-1}$ for cosmologically "interesting" axions.

Axion laboratory searches

Laboratory searches for axions cost $G_{a\gamma\gamma}^2$: you need to first *produce* an axion, and then *detect* it, typically via Primakoff-type processes like

$$\gamma + Ze \leftrightarrow Ze + a.$$

In a large-scale magnetic field $\vec{B}$, coherent over a length L, the process can be viewed also as an axion–photon oscillation, similar to neutrino flavor oscillations: the conversion probability will thus be proportional, as for neutrino oscillations, to a term of the type $\sin^2(\#m_a^2 L/\omega)$, where ω is the energy of the incident photon beam. If the argument is $m_a^2 L/\omega \ll 2\pi$, the axion beam is "coherent" with the incident photon beam, and the conversion probability reads [183]

$$\Pi \sim \frac{1}{4} G_{a\gamma\gamma}^2 B^2 L^2.$$

One setup is to shine a laser beam through a superconducting magnet to produce axions; put an optical barrier (wall); put a second magnet and a detector to see if any photons are regenerated. The probability of detecting the second photon is then proportional to Π^2. If "light shines through the wall" it might be because of axions!

Alternately, one can use the effect of *lost* photons on the beam itself, whose polarization would be affected passing through a magnetic field and producing axions. In fact, one such effect (dichroism) was detected by the PVLAS collaboration [184], but not confirmed by other experiments and attributed to instrumental effects [185].

Microwave cavities

A less costly (in terms of powers of $G_{a\gamma\gamma}$) approach is to try to convert the Galactic axions in the halo into quasi-monochromatic photons. For cosmologically interesting axions ($m_a \sim 10\mu$eV) this corresponds to microwave frequencies (roughly, 2 GHz, as we estimated above). Now, there are a lot of $m_a \sim 10\mu$eV axions around, if they are indeed the dark matter, as you will show in the following exercise:

> **Exercise 126.** Assume a local dark matter density $\rho_{\rm DM} \sim 0.3\,{\rm GeV/cm^3}$. Show that this corresponds to a spectacular number density of axions,
>
> $$n_a \sim 3 \times 10^{13}/{\rm cm}^3,$$

(Continued)

Exercise 126. (*Continued*)

and thus to a flux, for $v \sim 10^{-3}c$, of

$$n_a v \sim 10^{21}/(\text{cm}^2\text{s}).$$

The coupling of a cavity to Galactic halo axions is proportional to

$$G_{a\gamma\gamma} a \vec{B} \cdot \int_{\text{Vol}} \vec{E} d^3 x.$$

The power transferred to the cavity from Galactic axions conversion is then [186]

$$P \sim 10^{-22} \text{W} \left(\frac{\text{Vol}}{10\,\text{l}}\right) \left(\frac{B}{6\,\text{T}}\right)^2 \left(\frac{m_a}{10\,\mu\text{eV}}\right).$$

Exercise 127. Try to derive the order of magnitude of the power above.

Such power is puny, but with large enough cavities, large enough magnetic fields ($B \sim 8$ T), and smart enough experimentalists to suppress the apparatus' noise, it is enough to set constraints probing the theoretically expected range for $G_{a\gamma\gamma}$ in the few μeV mass range, and, hopefully in the near future, extending the sensitivity to much wider mass range [187].

Solar axions

Another technique that only costs $\mathcal{O}(G_{a\gamma\gamma})$ is looking for axions produced by Primakoff processes[1] in the Sun. These axions have an energy $E_a \sim 4.2$ keV and, as discussed above in Sec. 7.2, the corresponding "axion luminosity" is

$$L_A \simeq 1.85 \times 10^{-3} L_\odot \left(\frac{G_{a\gamma\gamma}}{10^{-10}\,\text{GeV}^{-1}}\right)^2,$$

[1]Historically, Primakoff processes are $A + \gamma \to A + \pi^0$, where N really represents the electromagnetic external field of an atom; here, the pseudoscalar particle is the axion instead of the π^0.

giving an X-ray flux at Earth of

$$3.75 \times 10^{11} \text{cm}^{-2}\text{s}^{-1} \left(\frac{G_{a\gamma\gamma}}{10^{-10}\ \text{GeV}^{-1}} \right)^2.$$

Exercise 128. Convert the luminosity above into the flux, please.

Now, detecting those axions is a bit trickier than detecting Galactic axions, since the coherence condition $|q|L \ll 1$, where the momentum transfer

$$|q| \equiv \frac{m_a^2}{2E_a} \sim 10^{-4}\text{eV} \left(\frac{m_a}{1\ \text{eV}} \right)^2,$$

implies $L \ll 1$ mm. That is a very impractical length-scale for laboratory magnets. However, not all is lost: in a plasma, photons acquire a "plasma mass" (as you might recollect from Kindergarten E&M)

$$m_\gamma \sim \omega_{\text{plasma}} \approx \sqrt{\frac{4\pi\alpha N_e}{m_e}} \simeq 30 \sqrt{\frac{Z}{A} \frac{\rho_{\text{gas}}}{\text{g/cm}^3}},$$

with N_e the plasma electron density and ρ_{gas} the gas density. The resulting momentum transfer is now

$$|q| = \frac{m_a^2 - m_\gamma^2}{2E},$$

thus by tuning the plasma mass, i.e. the gas density, to satisfy $m_a \sim m_\gamma$ these "axion telescope" can be made to detect axion of a specific mass quite efficiently: the conversion probability will then scale as $P \sim B^2 L^2$, and the maximal X-ray flux produced by solar axions is

$$\phi_{\text{max}} \sim 10^{-8} \frac{1}{\text{cm}^2\text{s}} \left(\frac{B}{3\ \text{T}} \right)^2 \left(\frac{L}{4\ \text{m}} \right)^2 \left(\frac{m_a}{1\ \text{eV}} \right)^4.$$

The best sensitivity obtained so far comes from the CERN Axion Solar Telescope (CAST) which cleverly recycled a decommissioned Large Hadron Collider (LHC) magnet with bores filled by gas (He^4 and He^3) at varying pressure. Such limits get close to theoretically interesting values of $G_{a\gamma\gamma}$ for $m_a \lesssim 1$ eV.

Chapter 8

Sterile Neutrinos as Dark Matter Particles

8.1 Praeludium

The discovery of neutrino masses and mixing indicates new physics beyond the Standard Model (SM): in the SM, in fact, only $SU(2)_L$ "left-handed" neutrinos exist, and they are strictly massless; in addition, they are both mass- *and* interaction-eigenstates, i.e. there can be no flavor mixing or oscillations). Having to build a new piece of the SM to address this shortcoming comes with the opportunity to "kill two birds with one stone"[a]: neutrino masses and mixing *plus* dark matter. Actually, in this case one even has the opportunity to kill a *third* bird: providing an explanation for the baryon asymmetry! Let us see how.

Barring postulating "weirdos" such as $SU(2)_L$ triplet "Higgs" bosons [188], the simplest addition to the SM cookbook needed to explain neutrino masses and mixing is a set of n fermions N_a ($a = 1, \ldots, n$) singlet ("sterile") under all SM gauge interactions. The neutrino sector, in addition to the old-fashioned $SU(2)_L$ "active" neutrinos ν_α ($\alpha = 1, \ldots, 3$), is described by the so-called see–saw Lagrangian

$$\mathcal{L} = \mathcal{L}_{\rm SM} + i\bar{N}_a \partial\!\!\!/ N_a - y_{\alpha a} H^\dagger \bar{L}_\alpha N_a - \frac{M_a}{2} \bar{N}_a^c N_a, \tag{8.1}$$

where the last term is a Majorana mass term for the "sterile" neutrinos (prohibited by gauge invariance for the active neutrinos, of course). Note that since the sterile neutrinos are, ahem, sterile, the Majorana mass matrix

[a]Or, as we say in Italian, "prendere due piccioni con una fava", which is the more birds-friendly "catch two pigeons with one fava bean".

can be chosen, without loss of generality, to be diagonal. The resulting mass eigenstates ν_i^m ($m = 1, \ldots, n+3$) are linear combinations of the active ν_α and of the sterile N_a neutrinos, resulting from diagonalizing the $(n+3) \times (n+3)$ mass matrix

$$M^{(n+3)} = \begin{pmatrix} 0 & y_{\alpha a}\langle H\rangle \\ y_{\alpha a}\langle H\rangle & \mathrm{diag}(M_1, \ldots, M_n) \end{pmatrix}.$$

Baby linear algebra will convince you that, as long as

$$y_{\alpha a}\langle H\rangle \sim yv \ll M_a \sim M,$$

i.e. as long as the mass scale of the sterile, right-handed neutrinos is much larger than the electroweak scale *times* the active neutrino Yukawa couplings, the eigenvalues split into (mostly) active, light neutrinos of mass

$$M(\nu_{1,2,3}) \sim \frac{y^2 v^2}{M},$$

and (mostly) sterile, heavier neutrinos of mass

$$m(\nu_a) \sim M.$$

Also, the see–saw Lagrangian will produce mixing angles of order

$$\theta^2_{\alpha a} \sim \frac{y^2_{\alpha a} v^2}{M^2},$$

between the active and the sterile neutrinos.

The key ingredient to explain the smallness of the mostly[b] active neutrino masses compared to the electroweak scale[c] is that the ratio $y^2v^2/M^2 \ll 1$. This, in turn, might either signal very small $y \ll 1$ and $v \gtrsim M$, or $y \sim 1$ and $M \gg v$. The latter possibility (where the light active neutrino masses are light because of the heaviness of their sterile, right-handed counterparts) is known as see–saw: the active neutrinos are light because the sterile neutrinos, on the other end of the see–saw, are heavy.

It is important to note that, in general, the scale M has *nothing to do* with the electroweak scale. Actually, one can come up with arguments in favor of a parametrically *small* M based on symmetry: 't Hooft's "naturalness" criterion [189] predicates that a theory parameter can be *naturally* (or *technically*) *small* if setting it to zero *increases the symmetry* of the Lagrangian, which would be the case here for $M \to 0$.

---

[b] An adverb we will drop hereafter.
[c] Some of the latest limits from cosmology indicate $\sum_\nu m_\nu < 0.43$ eV.

Exercise 129. Argue that for $M \to 0$ indeed the symmetry of the SM plus sterile neutrino Lagrangian above is increased.

We will see later that one possibility is actually that the scale M originates from the vacuum expectation value of a real singlet scalar S, which might be associated with the electroweak scale and with a non-thermal production mechanism for sterile neutrinos in the early universe.

In this chapter we are interested in models where the lightest of the "sterile" neutrinos is the dark matter. Effectively, this possibility corners a relatively small range of masses (around the keV scale) and mixing angles. Excitingly, since these sterile neutrinos are by construction meta-stable and possess a one-loop decay mode into one active neutrino and a quasi-monochromatic photon, there exists a possibility of discovering sterile neutrinos with astronomical observations in the frequency range corresponding to that mass range: X-rays.

8.2 Sterile neutrino dark matter?

If we are to contemplate sterile neutrinos as dark matter particles, we must first make sure that they are long-lived enough, i.e. at least longer-lived than the universe, $\tau_U \sim 10^{17}$ s. No symmetry protects sterile neutrinos from decaying into (3) active neutrinos.

Exercise 130. Explain why 3 is the smallest number of active neutrinos a sterile neutrino can decay into.

From the discussion above we learned that the generic sterile neutrino coupling to matter is of order θG_F, since it originates exclusively from the mixing (of $\mathcal{O}(\theta)$) with active neutrinos (which in the low-energy regime interact with matter through an effective four-fermion interaction with effective coupling G_F, see Appendix A.3). If I were to estimate the lifetime of a sterile neutrino with a mixing angle θ and mass m I would write

$$\Gamma \sim \theta^2 G_F^2 m^5 \sim \theta^2 \left(\frac{m}{\text{keV}}\right)^5 10^{-40}\,\text{GeV} \Rightarrow \tau \sim 10^{16}\text{s}\,\theta^{-2}\left(\frac{m}{\text{keV}}\right)^{-5},$$

matching the mass dimensions of Γ with powers of the only mass scale in the problem, m (and assuming active neutrinos are effectively massless).

Exercise 131. Cross check the 10^{16} s figure above.

Exercise 132. Estimate the largest possible sterile neutrino mass for the particle to be stable on cosmological scales (i.e. $\tau \gtrsim \tau_U$).

We thus would want the product

$$\theta^{-2}\left(\frac{m}{\text{keV}}\right)^{-5} \gg 1,$$

which singles out a corner of the parameter space we will mostly be concerned with in this chapter, the (θ, m) plane: Roughly, to be long-lived enough sterile neutrinos must be lighter than $m \ll 1\ \text{keV}/\theta^{2/5}$.

Exercise 133. The detailed calculation of the sterile neutrino decay into three neutrinos is not complicated. Carry it out and show that

$$\Gamma_{3\nu} = \frac{G_F^2 m^5 \theta^2}{96\pi^3} \qquad \Rightarrow \qquad \tau_{3\nu} \simeq 10^{14}\ \text{years}\left(\frac{10\ \text{keV}}{m}\right)^5\left(\frac{10^{-8}}{\theta^2}\right).$$

Unlike for axions and other bosons, light-ish fermion dark matter candidates (as, potentially, sterile neutrinos) must obey generic constraints from Fermi statistics, which we described in Sec. 1.9. A more general limit, known as the *Tremaine–Gunn bound* [43], stems from the fact that for classical collisionless fluids the maximum value of the phase-space density f_{max} must *decrease* with time. This fact does not depend on the spin statistics of the particle, and can be applied to both fermions and bosons. Suppose you consider an initial, homogeneous maximal phase-space density

$$f_{\text{initial}} = \frac{2}{h^3},$$

and compare it with the largest *observed* phase-space density, for example that of a dense galaxy today. Take, for the latter, following the original Ref. [43], an isothermal sphere with a central density ρ_0, and a one-dimensional Maxwellian velocity distribution with velocity dispersion

σ for the *final* phase-space density distribution. The maximal phase-space density is then

$$f_{\max} = \rho_0 m^{-4} (2\pi\sigma^2)^{-3/2}.$$

Exercise 134. Prove the equation above.

Defining the isothermal sphere's core radius r_c as

$$r_c^2 = \frac{9\sigma^2}{4\pi G_N \rho_0},$$

the requirement that the maximum phase-space density has *decreased* gives

$$m^4 > \frac{9h^3}{4(2\pi)^{5/2} G_N \sigma r_c^2}, \tag{8.2}$$

Exercise 135. Prove Eq. (8.2).

Exercise 136. Use typical Galactic values for $r_c \sim 1$ kpc and $\sigma \sim 100$ km/s to show that the limit on m from Eq. (8.2) is

$$m \gtrsim 100 \text{ eV}.$$

Keep in mind that limits obtained along similar lines to the Tremaine–Gunn limit depend on how sterile neutrinos are produced, since this affects the initial phase-space density distribution, and as a result there are ways around the limit based on different production mechanisms. Roughly speaking, though, we can keep in mind $m \gtrsim 1$ keV as a benchmark.

8.3 Sterile neutrino production

Sterile neutrinos are *not* eigenstates of the neutrino mass matrix, and they therefore mix with their active counterparts. Suppose for simplicity that all of the mixing takes place exclusively between the electron neutrino ν_e and the sterile (right-handed) neutrino N_1, and let us call the mass eigenstates of this two-particle system $\nu_{1,2}$; then the linear superpositions that give the mass eigenstates in terms of the interaction eigenstates are

$$|\nu_1\rangle = \cos\theta|\nu_e\rangle - \sin\theta|N_1\rangle,$$
$$|\nu_2\rangle = \sin\theta|\nu_e\rangle + \cos\theta|N_1\rangle,$$

with $\theta \ll 1$ and thus ν_2 corresponding to our sterile neutrino dark matter candidate ν_s. Dodelson and Widrow were the first to calculate the relic abundance of sterile neutrinos produced by oscillations from the active to the sterile neutrinos in the early universe [190]. Due to their small coupling to the thermal bath, sterile neutrinos are produced out of equilibrium, and simply accumulate over time. Before getting to the subtleties and complications of this calculation, let us estimate the temperature at which the rate of production in the early universe via coupling to the active neutrinos is maximal. The co-moving number density of produced sterile neutrinos at a given temperature is approximately given by the product of the production rate

$$\Gamma_{\nu_s} \sim (G_F^2 T^2)\theta^2(T) \cdot T^3,$$

itself the product of the production cross section times the number density ($\sim T^3$), times the Hubble time (i.e. inverse Hubble rate). The key complication stems from the fact that the mixing angle actually depends on temperature, as indicated explicitly above. Specifically, matter and quantum damping suppress oscillations, producing an effective mixing angle in matter θ_M that reads [191]

$$\theta \to \theta_M \simeq \frac{\theta}{1 + 2.4\left(\frac{T}{200\ \text{MeV}}\right)^6 \left(\frac{1\ \text{keV}}{m}\right)^2}. \tag{8.3}$$

As a result, at large temperatures $T \gg 200$ keV the production rate is strongly suppressed, $\Gamma_{\nu_s}(T \gg 200\ \text{MeV}) \sim T^{-7}$, thus peak production (i.e. the largest $\Gamma_{\nu_s}(M_P/T^2)$) will happen at a temperature

$$T_{\text{peak}} \sim 170 \left(\frac{m}{1\ \text{keV}}\right)^{1/4}\ \text{MeV}.$$

Exercise 137. Calculate T_{peak} and verify the assertion above.

The complete calculation would give roughly similar numbers [191]. Most of the action then takes place around the quantum chromodynamics (QCD) phase transition, when a large change in the number of effective degrees of freedom occurs. An approximation to the numerical results for the production of sterile neutrinos through oscillation is [192]

$$\Omega_{\nu_s} h^2 \sim 0.1 \left(\frac{\theta^2}{3 \times 10^{-9}} \right) \left(\frac{m_s}{3\ \text{keV}} \right)^{1.8} .$$

The biggest change to this calculation happens in the presence of a lepton asymmetry, which produces large enhancements due to the Mikheyev–Smirnov–Wolfenstein (MSW) effect [193, 194], especially at small mixing angles (the so-called Shi–Fuller resonant production mechanism [195]). In practice, the denominator in Eq. (8.3) can go to zero for certain momentum values, such that the asymmetric neutrino or antineutrino population gets transferred effectively to the population of sterile neutrinos. The resulting population of resonantly-produced *sterile* neutrinos is bound from above by the absolute value of the lepton asymmetry, $n_{\text{res}} \lesssim |n_\nu - n_{\bar{\nu}}|$, and for keV sterile neutrinos this mechanism is of interest (i.e. competitive with the abundance of non-resonantly produced sterile neutrinos) for typical lepton asymmetries larger than $\sim 10^{-6}$ [191].

Exercise 138. Calculate the region in the (m, θ) parameter space such that sterile neutrinos are in thermal equilibrium at some point in the early universe. After finishing reading this chapter, argue that sterile neutrinos compatible with Lyman-α, dSph and X-ray constraints have never been in thermal equilibrium in the early universe. As a result, bounds such as the Lee–Weinberg [52] limit obviously do not apply!

In the context of resonant production of sterile neutrinos, Laine and Shapshnikov [196] have pointed out an explicit model where one could indeed kill *three* birds with one stone (or let them live, and feed them with a single fava bean, in Italian). The scenario, known as minimal SM with neutrino masses, or νMSM, posits that the minimal set of ingredients to explain (i) neutrino masses and oscillations, (ii) dark matter, and (iii) the baryon asymmetry in the Universe is to postulate three sterile neutrinos, one of them, the lighter, with a mass M_1 in the keV range responsible for the dark matter via the Shi–Fuller resonant production mechanism, and the two more massive sterile neutrinos $M_2 \simeq M_3 \sim 1 - 10$ GeV, with $|M_2 - M_3| \sim 1$ keV, responsible for the dynamical generation of a lepton asymmetry L which is then converted to a baryon asymmetry $B \sim L$. Here, neutrino oscillations, instead of heavy neutrino decay, are responsible for the generation of $L \neq 0$. As long as $M_2 \sim M_3$, CP-violating effects in the oscillations of active neutrinos are amplified, producing a large enough L. The price of minimality is some degree of fine-tuning, $\Delta M_{23}/M_{2,3} \sim 10^{-6}$, which might perhaps be addressed positing symmetry arguments [197].

Another interesting and suggestive mechanism for the production of relic sterile neutrinos is via the decay of a singlet, real scalar S coupled to right-handed neutrinos via a coupling term of the form

$$\frac{h_a}{2} S \bar{N}_a^c N_a,$$

and to the SU(2) Higgs via a standard "Higgs portal" type coupling

$$SH^{\dagger}H \text{ and/or } S^2 H^{\dagger} H,$$

in the scalar potential $V(H, S)$. The sterile neutrinos are produced via the $S \to NN$ decay, and are "cooled" by the drop in the number of effective degrees of freedom from the $g_*(T_S \sim 110 \text{ GeV}) \sim 110$ down to $g_*(T \sim 0.1 \text{ MeV}) \sim 3.36$, making this population of non-thermally produced sterile neutrino much colder than those produced via oscillation. Additionally, if V produces a non-zero vacuum expectation value for S, then the Majorana mass term for the right-handed neutrinos is dynamically generated, with $M \sim h\langle S \rangle$, the symbol $\langle S \rangle$ indicating the vacuum expectation value for S [192].

Let us put in some numbers. Assume that the potential V drives the mass $M_S \sim 10^2$ GeV. The decay rate for S is given by

$$\Gamma_{S \to NN} \sim \frac{h^2}{16\pi} M_S.$$

To estimate the number density of sterile neutrinos produced by S decays, we multiply the decay rate by (1) the S number density and (2) the timescale for the decay (i.e. the inverse Hubble rate at the time of decay). Now, the longest time scale at which S are still abundant (i.e. relativistic) is at $T \sim M_S$, and $n_S \sim T^3$. The estimate for the co-moving N number density is thus

$$\frac{n_N}{s} \sim \frac{n_S}{s} \tau \Gamma \sim \frac{M_P}{M_S^2} \frac{h^2}{16\pi} M_S.$$

Let us now calculate

$$\Omega_N = M_S \left(\frac{n_S}{s} \right)_{M_S} s_0 \sim \frac{h^2}{16\pi} \frac{h\langle S \rangle}{M_S} M_P \frac{g_*(100 \text{ GeV})}{g_*(0.1 \text{ MeV})},$$

with the ratio $\xi = \frac{g_*(100\text{ GeV})}{g_*(0.1\text{ MeV})}$ accounting for the change in the effective degrees of freedom. Plugging in numbers we get

$$\Omega_N \sim 0.2 \left(\frac{h}{10^{-8}}\right)^3 \frac{\langle S \rangle}{m_S}.$$

This is very interesting, because if $\langle S \rangle \sim 100-1000$ GeV, and $h \sim 10^{-8}$, the sterile neutrino mass $m \sim h\langle S \rangle \sim$ few keV, which is the cosmologically interesting value we would indeed like to see: much lighter masses would die because of Tremain–Gunn-type bounds, and much heavier masses would make for too-fast decaying dark matter. Also, the average sterile neutrino dark matter momentum is suppressed (i.e. these non-thermal sterile neutrinos are colder) which as we shall see in Sec. 8.4 has important implications for structure formation [192].

8.4 Sterile neutrino dark matter, and how to discover it

So far we convinced ourselves that (i) sterile neutrinos have plausible reasons to exist, independent of the dark matter problem, and that (ii) for some combination of the relevant (m, θ) parameters they are "good" particle dark matter candidates. The next question is: do sterile neutrinos have distinctive features as dark matter candidates? Are there clear-cut ways we can hope to detect them with?

A first distinctive feature of sterile neutrinos is that they are potentially "warm" dark matter candidates: what that means is that particles are produced relativistically, but their momenta are then redshifted to non-relativistic values in the radiation domination era. Structures form similarly to cold dark matter cosmologies at distances larger than some characteristic free-streaming scale, a scale under which particles stream out of overdense regions and erase density fluctuations. The co-moving free-streaming scale can be estimated, as a function of temperature, as [198]

$$\lambda_{\rm FS} \sim 1\text{ Mpc}\left(\frac{1\text{ keV}}{m}\right)\frac{\langle p_N \rangle}{\langle p_\nu \rangle},$$

where $\langle \cdot \rangle$ are average momenta as a function of temperature for the active (p_ν) and sterile (p_N) neutrinos, respectively, and the ratio $\langle p_N \rangle / \langle p_\nu \rangle$ accounts for sterile neutrinos being produced non-thermally [198].

Exercise 139. Prove the equation above, using arguments similar to what we talked about in Sec. 3.6.

If the sterile neutrino population is non-relativistic, then the corresponding free-streaming length is affected by the ratio $\langle p_N \rangle / \langle p_\nu \rangle$. In practice, that ratio is of order 0.9 for production off-resonance, 0.6 for resonant production à la Shi–Fuller, and 0.2 for production at $T \sim 100\,\text{GeV}$ [192]. The corresponding mass inside λ_{FS} under which structure formation is highly suppressed reads

$$M_{\text{FS}} \sim 3 \times 10^7\, M_\odot \left(\frac{10\,\text{keV}}{m} \right)^3 \left(\frac{\langle p_N \rangle}{\langle p_\nu \rangle} \right)^3,$$

Exercise 140. Calculate M_{FS} from λ_{FS}.

Note that for $\lambda \gtrsim \lambda_{\text{FS}}$ structure formation for sterile neutrino dark matter and for the standard cold dark matter is virtually indistinguishable. On the other hand, one can use the (more or less directly) observed small scale "abundance" to set constraints on m. In practice this is done using the observed Lyman-α absorption lines in the spectra of distant quasar, produced by neutral hydrogen clouds along the line of sight. This is no easy business, as the constraints depend both on hydrodynamical simulations affected by potentially large and hard-to-quantify systematics, and on assumptions on cosmological parameters. Roughly, for sterile neutrinos produced in the standard oscillation scenario, the limits are in the $m \gtrsim 2$–$4\,\text{keV}$ range [199–201]; Most recent data from 13,000 (!) quasar spectra from the BOSS survey indicate that, at the 2σ level, the limit is $m \gtrsim 4.35\,\text{keV}$ [202].

The scale M_{FS} for masses in the few keV range is quite suggestive, as it is close to the mass scale where cold dark matter cosmologies typically overproduce structure. In addition, that scale is close to the scale indicated by some analyses as being the smallest scale at which dark matter "clumps" are observed. While the jury is still out as to whether or not there exists a "small-scale problem" in ΛCDM, it is interesting to keep in mind that sterile neutrinos, depending on the particular model incarnation, might well have something to do with such small-scale issues.

Sterile neutrinos and pulsar kicks

The observed distribution in the velocity of pulsars has an anomalous high-velocity tail that is hard to explain in standard supernova explosion models: while the average velocity is estimated to be between 250 and 500 km/s, 15% of pulsars are observed to have velocities greater than 1000 km/s , with some as fast as 1600 km/s. Since most of the total supernova energy, up to 99% of the $\sim 3 \times 10^{53}$ erg is radiated via the emission of neutrinos, it is not unreasonable to ask whether, in the presence of sterile neutrinos, the high-velocity pulsars problem might find an explanation.

The total momentum carried out of the supernova by neutrinos is of the order of

$$p_{\nu,\text{total}} \sim E/C \sim 10^{43}\, \text{g}\frac{\text{cm}}{\text{s}},$$

to be compared with the momentum p_* of a neutron star of mass $\approx 1.4\, M_\odot$ and speed $v = 1000$ km/s,

$$\begin{aligned} p_* &= (1.4\, M_\odot)v \approx 3 \times 10^{41} \left(\frac{v}{1000\ \text{km/s}}\right) \text{g}\frac{\text{cm}}{\text{s}} \\ &\approx 0.03 \left(\frac{v}{1000\ \text{km/s}}\right) p_{\nu,\text{total}}, \end{aligned}$$

indicating that a few percent asymmetry in the neutrino emission could give enough of a "kick" to explain the anomalously large observed velocities.

Neutron stars have very large magnetic fields, and have large enough densities that active neutrinos undergo multiple scattering through processes (known as *urca* processes[d]) of the form

$$\begin{aligned} n + e^+ &\leftrightarrow p + \bar{\nu}_e, \\ p + e^- &\leftrightarrow n + \nu_e. \end{aligned}$$

[d]The origin of this denomination is amusing. Apparently George Gamow and Mario Schenberg were discussing these processes while in a Rio de Janeiro casino called *Cassino da Urca* when Schenberg declared that "the energy disappears in the nucleus of the supernova as quickly as the money disappeared at that roulette table". Interestingly, *urca* means robber in Gamow's south Russian dialect. It means "wow!" in Northern Italian instead.

Exercise 141. Show that for a neutron star of density $\rho_* \approx 5 \times 10^{17}$ kg/m^3 MeV neutrinos indeed undergo multiple scattering, by calculating the ratio of the mean free path for the process, $\lambda \sim 1/(\sigma \cdot \rho_*)$ and comparing with the typical radius of a neutron star, $R_* \sim 13$ km.

As a result, active neutrinos are *always* asymmetric, since the electrons, neutrons, and protons are polarized by the magnetic field and the cross section for the urca processes depend on the relative orientation of the neutrino momentum with respect to the electron spin,

$$\sigma(e^- \uparrow, \nu \uparrow) \neq \sigma(e^- \uparrow, \nu \downarrow).$$

However, in the presence of multiple scattering neutrinos remain in equilibrium as they diffuse through the neutron star, and no net asymmetry can be generated as the neutrinos escape the star.

Things of course change if the *active* neutrinos turn into *sterile* neutrinos inside the (proto-)neutron star: sterile neutrinos have a much longer mean free path, i.e. after they are produced they exit the neutron star without interacting, effectively preserving the asymmetry created in the urca processes (as you will convince yourself of with the following exercise).

Exercise 142. Estimate the mean free path of sterile neutrinos of mass m and mixing angle θ inside a neutron star and compare to the typical neutron star size.

The asymmetry in the momentum distribution from sterile neutrino production depends on whether there is or not an MSW resonance and, if so, whether the resonance takes place inside or outside the neutron star core. For example, for an MSW resonance *outside* the core, Ref. [192] calculates a momentum distribution asymmetry on the order of

$$\frac{\Delta k_s}{k} = 0.03 \left(\frac{T}{3\ \mathrm{MeV}}\right)^2 \left(\frac{\mu_e}{50\ \mathrm{MeV}}\right) \left(\frac{h_e}{6\ \mathrm{km}}\right) \left(\frac{B}{10^{15}\ \mathrm{G}}\right),$$

large enough, for typical values of the parameters (μ_e being the electron chemical potential and $h_e = [d(\ln N_e)/dr]^{-1}$ the scale height of the electron density), to explain the anomalously large velocities of certain pulsars.

Interestingly, if indeed sterile neutrinos are responsible for pulsar kicks, there might be an observable consequence: gravitational waves. The magnetic field is not aligned with the axis of rotation, so the outgoing neutrinos create a non-isotropic source for gravitational waves. The resulting gravity wave signal, for a nearby pulsar (say, within 1 kpc) would be observable by LISA or even advanced LIGO [203].

Exercise 143. Estimate the maximal energy density you would expect to be associated with a gravity wave signal from such a nearby pulsar.

X-rays from sterile neutrinos

While the dominant sterile neutrino decay mode is into three neutrinos, there is a one-loop process that leads to the production of quasi-monochromatic photons: $\nu_s \to \gamma\nu_a$. If I were to guess-estimate the decay width for such a process I would probably write

$$\Gamma_{\nu_s \to \gamma\nu_a} \approx \frac{\alpha}{16\pi^2}\theta^2 G_F^2 m^5.$$

Exercise 144. Explain the reasoning behind the factors of α, $16\pi^2$ and the rest of the equation above.

The estimate above would have overestimated the actual decay width, but not by too much (I assume $\theta \ll 1$):

$$\Gamma_{\nu_s \to \gamma\nu_a} = \frac{9\alpha}{256\pi^4}\theta^2 G_F^2 m^5 \simeq \frac{\theta^2}{1.8 \times 10^{21}\ \mathrm{s}}\left(\frac{m}{1\ \mathrm{keV}}\right)^5,$$

meaning that (1) this is a very slow decay mode, and (2) $\Gamma_{3\nu}/\Gamma_{\gamma\nu} \simeq 130$, more or less what one would expect.

Being a two-body final state, and since $m = m_{\nu_s} \gg m_{\nu_a}$, the photon energy essentially corresponds to $E_\gamma \simeq m/2$. For keV sterile neutrinos this is an X-ray line, with a width associated with the virial motion of dark matter particles, between $\Delta E/E \sim \sigma_v/c \sim 10^{-4}$ for dwarf galaxies and $\Delta E/E \sim 10^{-2}$ for clusters. The flux of X-ray photons from a given direction ψ received by an instrument of a given angular field of view (fov) is

$$\phi_\gamma = \frac{\Gamma_{\gamma\nu}}{4\pi}\frac{E_\gamma}{m}\int_{fov} d\Omega \int_{\text{line of sight}} \frac{\rho_{\rm DM}}{m} dr(\psi) = \frac{\Gamma_{\gamma\nu}}{8\pi m} J(\Delta\Omega, \psi),$$

where the field of view spans a solid angle $\Delta\Omega$, the 8π in the final equation comes from $E_\gamma/m \simeq 1/2$, and you recognize the $J(\Delta\Omega\psi)$ we talked about in Sec. 5.6 — this time, though, the fields of view of typical X-ray telescopes are much smaller than those we considered before for gamma-ray telescopes, on the order of a few arcmin.

Interestingly, for a variety of astrophysical objects from dwarf galaxies to clusters, the resulting J factors within, say, a cone of radius 6 arcmin (roughly the correct figure for telescopes such as XMM or Chandra) is similar. Up to variations in the details of the dark matter density profiles, I find that for nearby clusters (e.g. Coma), for M31, for the Galactic center or for local dwarf spheroidal galaxies (e.g. Draco), the J's are all between a few $\times 10^{17}$ and a few $\times 10^{18}$ GeV/cm^2. Which object is the best bet? And which constraints can we expect to derive?

Clusters and, to a lesser extent, galaxies such as the Milky Way and M31 host thermal gas that radiates thermal bremsstrahlung in the few keV (clusters) or in the fraction of a keV (galaxies) range. This is one annoying background dSph are not affected by, being essentially devoid of any significant amount of interstellar gas. And of course there is point-source subtraction, especially bothersome in the case of powerful central AGN. But to run some numbers, let us use the unresolved diffuse cosmic X-ray background (CXB), which is there in all cases, and is at the level of

$$\phi_{\rm CXB} \sim 9.2 \times 10^{-7} \left(\frac{E}{1\ \rm keV}\right)^{-0.4} \rm cm^{-2}\ s^{-1}\ arcmin^{-2} \ \rightarrow\ \sim 10^{-4}\ cm^{-2}\ s^{-1},$$

for the angular region of interest to us. Let us plug in numbers:

$$\phi_\gamma = \frac{\Gamma_{\gamma\nu}}{8\pi}\frac{J}{m} \sim 10^{-4}\ {\rm cm^{-2}\ s^{-1}} \left(\frac{\theta^2}{10^{-7}}\right)\left(\frac{m}{1\ \rm keV}\right)^4\left(\frac{J}{10^{18}\ \rm GeV/cm^2}\right),$$

meaning that we can estimate the X-ray constraints as

$$\left(\frac{\theta^2}{10^{-7}}\right)\left(\frac{m}{1\ \rm keV}\right)^4 \lesssim 1.$$

Perhaps surprisingly, this estimate turns out to be actually more or less correct (see e.g. Fig. 13 in [192]).

In the phenomenologically interesting regime of $m \sim$ few keV and $\theta^2 \sim 10^{-10}$ the X-ray signal from sterile neutrino decay can be detectable with current instruments, and some scholars claim it has been. Let me give a succinct account of this recent story, and of the associated, necessary burden of proof of getting *really "sure"* we are detecting dark matter.

In early 2014 two groups [204, 205] independently detected an X-ray line at around 3.55 keV from the spectra of stacked clusters of galaxy, from the individual Perseus cluster, and Ref. [205] from M31. Both groups discussed the possible contamination from elemental lines — de-excitation of partly ionized atoms in the intracluster medium or in the IGM, specifically a line[e] from a He-like Potassium ion, known to atomic physicists as K XVIII (the XVIII indicates that the potassium atom has lost 18-1 of its 19 electrons — thank you atomic physicists for the very user-friendly notation!).

Exercise 145. Estimate at which Z hydrogen-like ions have their 1s transition at about 3.55 keV; Try to correct for helium-like ions such as K XVIII (for which $Z_{\rm eff} \lesssim Z$ because of the second electron), and argue that indeed $Z \sim 19$ is reasonable.

Estimating the brightness of such lines is no easy task: they depend both on the intracluster medium temperature (possibly a linear superposition of multiple temperatures, in fact), and on the relative elemental abundances (one normalizes the estimates to known bright lines, using nominal relative abundances, in this case solar abundances). The conclusion was that the two relevant K XVIII lines are too faint by factors of ~20–30, for certain choices of temperature and using relative elemental abundances as observed in the Sun. If interpreted as a decaying sterile neutrino, the signal was compatible, at face value, with $m \simeq 7.1$ keV and $\theta^2 \simeq 1.8 \times 10^{-11}$.

My collaborator Tesla Jeltema at UCSC and I set out to seek confirmation for the exotic nature of the line and to set constraints on such nature, using archival XMM observations of the Galactic center region [206]. To our delight, we indeed observed a 3.5 keV line, with a brightness roughly in agreement with what expected from the sterile neutrino decay hypothesis of Refs. [204, 205]. However, we also pointed out that the brightness of the K XVIII line could well lie within a factor of around 3 of what predicted using solar abundances. Given that there is no reason for the Galactic center region to mirror relative elemental abundances as in the Sun, we argued that the K XVIII lines were a reasonable explanation to the 3.5 keV line. Also, we showed that the temperature chosen in the plasma models of Ref. [204] were biased towards large values, artificially suppressing (by up to one order of magnitude) the brightness of the K XVIII lines [207].

[e]Actually, a complex of lines in close proximity in energy.

We thus argued that an elemental origin was not at all ruled out as an explanation to the 3.5 keV line.

Later, it was also pointed out [208] that the potassium elemental abundance used in Refs. [204, 205] utilized *photospheric* values, instead of *coronal* abundances. Ref. [208] showed that this produced an underestimate in the relative abundance of potassium of a factor between 9 and 11, and led the authors of Ref. [208] to conclude that "there is therefore no need to invoke a sterile neutrino interpretation of the observed line feature at $\sim$3.5 keV".

Subsequent studies failed at finding any confirmation of the sterile neutrino decay origin of the 3.5 keV line in environments devoid of potential background from elemental de-excitation lines. Specifically, studies targeted stacks of low-mass groups and galaxies [209], local dwarf spheroidal galaxies [210] without finding any hint of a signal at 3.5 keV. Other studies used cluster observations with other X-ray telescopes, specifically Suzaku, to confirm our claim that the K XVIII could indeed explain the 3.5 keV line where observed [211]. The XMM satellite was subsequently used for an ultra-deep (longer than 1.5 Ms) observation of the nearby Draco dSph, a virtually background-free target, and again no signal was observed [212].[f] Finally, with Jeltema and Carlson at UCSC we showed that the *morphology*, i.e. the spatial distribution of the 3.5 keV photons from the Galactic center follows the observed morphology of other bright elemental lines, and that for the Perseus cluster the emission peaks around the cluster's cool core, which is markedly smaller than the dark matter halo scale radius, and which, again, correlates with where other bright elemental lines are observed [213].

At the moment, it appears that Occam's razor distinctly points towards a "boring" K XVIII origin for the 3.5 keV line. The sterile neutrino decay possibility is at face value excluded, both from morphology and from the non-detection from Draco. Other clever possibilities (such as axion-like particle (ALP) conversion in the presence of magnetic fields [214], or the excitation of dark matter to a heavier, unstable state by collisions with the thermal plasma [10]) remain perfectly viable both from the standpoint of the absence of a signal from small systems, and from that of morphology.

---

[f]The Referee for this paper was remarkably zealous: he or she let us know that "after I was asked to referee this paper, I decided to download the data and examine the spectrum myself. I largely agree with your conclusions regarding the absence of a notable feature at $\sim$3.5 keV, as well as your limits on the line flux in this region."

Up until March 2016, hope was that the successfully launched Hitomi (formerly known as Astro-H) satellite would settle the dust, with an instrument onboard featuring a high-enough energy resolution to not only distinguish the energy of the K XVIII lines (and the two lines themselves) from a line at a different energy, but also to resolve the line width and thus to point towards a thermal or a Doppler broadening, or a mix of the two as in the case of the thermally excited dark matter. Unfortunately, contact with the satellite was lost on March 26, 2016, and what are thought to be the remnants of the spacecraft were observed to tumble in orbit out of control. During its short life, Hitomi did observe the Perseus cluster, but found no evidence for a 3.5 keV line [292]. Such is life, sometimes.

Chapter 9

Bestiarium: A Short, Biased Compendium of Notable Dark Matter Particle Candidates and Models

9.1 Praeludium

What makes a dark matter particle candidate a *good* candidate? One criterion is that a dark matter candidate ought to at least kill more than one bird with one stone. You do not just postulate it exists *ad hoc* to solve the dark matter problem; you also tackle some other outstanding problem (you get to decide in which arena). For example, weakly interacting massive particles (WIMPs) are often connected to theories at the electroweak scale that offer solutions to things like the hierarchy problem, or gauge coupling unification, or both; sterile neutrinos might be connected with the question of why Standard Model (SM) "active" neutrinos have non-zero masses and mixing, and perhaps to the generation of a baryon asymmetry in the universe; and axions are motivated by a dynamical solution to the strong CP problem.

Tackling more than one open problem is not, however, a necessary condition for a dark matter candidate to be "well motivated". Perhaps the key is to add as little as possible to the particle physics cookbook: *minimality*, that is. How exactly one quantifies minimality in model building is largely subjective. However, this remains an interesting and widely employed guideline.

Of the vast bestiary of dark matter candidates that have been proposed over the years, there have been numerous proposals for "good" dark matter particles besides WIMPs, sterile neutrinos, and axions. Some of them are more than legitimately entitled to be mentioned and briefly described in a book which is a primer on particle dark matter.[a] My highly-biased selection of "odd beasts" in fact only boils down to gravitinos (Sec. 9.2), with a few quantum-field-theoretical details and some remarks on other super-weakly interacting dark matter candidates, and to super-heavy dark matter particles which are never in thermal equilibrium, sometimes called "WIMPzillas" (Sec. 9.3). The discussion on gravitinos touches on some important pieces of theory related to local supersymmetry (supergravity), and that on super-heavy dark matter on the interesting possibility of gravitational production of dark matter. I then briefly describe a couple of *classes* of dark matter models that have (legitimately) attracted significant attention recently: dark matter models with (significant) self-interactions (Sec. 9.4), and asymmetric dark matter (Sec. 9.5). Finally, Sec. 9.6 presents a few examples of models built with (some incarnation of) the notion of "*minimality*" in mind: variations on the theme of neutrinos, the real scalar singlet, the "original" $SU(2)_L$ minimal dark matter model, the inert doublet model, and dark photons.

9.2 Of gravitinos and other depressing dark matter candidates

Often, life is not easy. Sometimes it is impossibly difficult. The *macroscopic* features of dark matter do not guarantee at all any *microscopic* connection with SM particles besides gravity. In other words, as long as certain gross properties of a dark matter particle candidate are fulfilled (for example, it is slow enough to form halos of the right size and at the right epoch in the early universe), the dark matter particle could possibly be completely decoupled from the particles we know and love.

As a matter of fact, the first supersymmetric dark matter particle candidate ever proposed[b] (and in some sense the most natural) is a particle that has very, very weak interactions[c] with our beloved SM particles:

---

[a]It goes without saying that this is *not* a complete list, nor is it meant to be; In case *your* preferred dark matter candidate is not included here only two things should be blamed: my own taste, and my personal bias as to whether or not I think there are unique and interesting physics lessons to be learned.

[b]See Ref. [215], sec. V.B for a brief history of supersymmetric dark matter candidates.

[c]As we will see below, "transverse" gravitino modes have interactions suppressed by the Planck scale, while "longitudinal" modes by the supersymmetry-breaking scale.

the gravitino [216], the spin 3/2 superpartner of the graviton. Before digging into estimates of the salient gravitino properties (lifetime, relic abundance,...), it is sobering[d] to think a little bit about how we would describe a spin 3/2 particle in the language of quantum fields.

In the supersymmetric limit, massless gravitinos are vector-spinor fields $\psi^a_\mu(x)$, the so-called (Majorana) Rarita–Schwinger fields, whose Lagrangian looks like this:

$$\mathcal{L} = -\frac{1}{2}\varepsilon^{\mu\nu\rho\sigma}\bar{\psi}_\mu\gamma_5\gamma_\nu\partial_\rho\psi_\sigma,$$

a Lagrangian which is invariant under the following (supersymmetric) gauge transformations with parameter ε:

$$\delta\psi_\mu = -2\bar{M}_P\partial_\mu\varepsilon.$$

where $\bar{M}_P = M_P/(8\pi) \simeq 2.4 \times 10^{18}$ GeV.

Exercise 146. Check said invariance, please.

We will not belabor on how the gravitino gets a mass by "eating" the massless Goldstino, via a so-called super-Higgs mechanism,[e] and providing the gravitino with its longitudinal modes; we will just write down (following Ref. [217]) the gauge-invariant action for the Goldstino (χ)-gravitino system: et voilà!

$$\mathcal{L} = -\frac{1}{2}\varepsilon^{\mu\nu\rho\sigma}\bar{\psi}_\mu\gamma_5\gamma_\nu\partial_\rho\psi_\sigma + \frac{1}{2}\bar{\chi}\gamma^\mu\partial_\mu\chi$$
$$-m_{3/2}\left(\frac{1}{4}\bar{\psi}_\mu[\gamma_\mu,\gamma_\nu]\psi_\nu + \bar{\chi}\chi - \sqrt{\frac{3}{2}}\bar{\psi}_\mu\gamma^\mu\chi\right).$$

For an interesting derivation of where the Goldstino mass term comes from and how the Higgs- and super-Higgs mechanism differ see Refs. [218,219] (I thank Francesco D'Eramo for bringing these references to my attention!). The Lagrangian above is invariant under the following local field

---

[d]Or appetite-wetting, for those with a weakness for formal theory.

[e]You should be familiar with the "non-super" SM boson analog.

transformations

$$\delta\psi_\mu = -\bar{M}_P(2\partial_\mu \varepsilon + i m_{3/2}\gamma^\mu \varepsilon),$$
$$\delta\chi = \sqrt{2}F\epsilon,$$

as long as the gravitino mass $m_{3/2}$ and the SUSY breaking vev F (which has dimensions of mass squared) are related by

$$m_{3/2} = \frac{F}{\sqrt{3}\bar{M}_P}.$$

Exercise 147. As above, make sure you check said invariance please.

The interaction term with matter (which one derives by defining a Majorana super-current S_μ from the functional variation of the matter action, and by demanding that the full action be invariant to zeroth order in $\bar{M}_P^{-1}$ — for gory details see e.g. [217]) is also forced to a fixed form by the requirement of invariance under local field transformations. After gauge fixing to the unitary gauge, the whole Lagrangian for the *massive* gravitino Ψ_μ can be cast as

$$\mathcal{L} = -\frac{1}{2}\varepsilon^{\mu\nu\rho\sigma}\bar{\Psi}_\mu\gamma_5\gamma_\nu\partial_\rho\Psi_\sigma - \frac{m_{3/2}}{4}\bar{\Psi}_\mu[\gamma^\mu, \gamma^\nu]\Psi_\nu + \frac{1}{2M_P}\bar{\Psi}_\mu S^\mu.$$

Exercise 148. Consider free gravitinos, $S^\mu = 0$. Show that their equation of motion is

$$-\frac{1}{2}\varepsilon^{\mu\nu\rho\sigma}\gamma_5\gamma_\nu\partial_\rho\Psi_\sigma - \frac{m_{3/2}}{4}[\gamma^\mu, \gamma^\nu]\Psi_\nu = 0,$$

an equation known as "Rarita–Schwinger" equation. Now use

$$\gamma^\mu\Psi_\mu(x) = 0, \quad \partial^\mu\psi_\mu(x) = 0,$$

to show that the Rarita–Schwinger equation reduces to a Dirac equation for Ψ_μ,

$$(i\gamma^\mu\partial_\mu - m_{3/2})\Psi_\mu(x) = 0.$$

Here, as anywhere else in this book, we are interested in the phenomenology of gravitinos in connection with the particle nature of dark matter. One possibility, which is essentially automatically realized in gauge-mediated supersymmetry breaking models (for a review see [220]), is that the gravitino is the LSP. In this case, if R parity is conserved, the gravitino is stable.

If gravitinos are much lighter than the typical mass scale M_s of the other supersymmetric particles (as, again, is typically the case in gauge mediation), the Goldstino-inherited longitudinal modes have an enhanced (by the ratio $M_s/m_{3/2}$, thus potentially by many orders of magnitude for keV-scale gravitinos!) interaction strength, making gravitinos effectively two-component Majorana fermions. For light gravitinos, say well below the electroweak scale, the relevant pair-annihilation cross sections into two photons and a fermion–antifermion pair are, as a function of s, the center-of-mass energy squared [221] (I will hereafter indicate the gravitino as a particle, rather than a field, with the symbol $\tilde{G}$)

$$\sigma(\tilde{G}\tilde{G} \to \gamma\gamma) = \frac{1}{576\pi} \frac{M_{\tilde{\gamma}}^2 s^2}{(m_{3/2} M_P)^4},$$

$$\sigma(\tilde{G}\tilde{G} \to \bar{f}f) = \frac{g_f}{720\pi} \frac{s^3}{(m_{3/2} M_P)^4},$$

where the number of degrees of freedom $g_f = 4, 2$ for charged fermions and neutrinos, respectively, and where $M_{\tilde{\gamma}}$ is the photino mass (following Ref. [221], I neglect neutralino mixing).

Exercise 149. Try to justify from the discussion above the power counting in $m_{3/2}$, M_P and s that appears in the pair-annihilation cross sections above.

In the regime we are interested in $s \ll M_{\tilde{\gamma}}^2$, so the two-photon channel dominates, and that cross section is what we will use that to calculate the gravitino freeze-out temperature

Exercise 150. Show that, neglecting factors of order 1, the gravitino freeze-out temperature for light gravitinos is

$$T_{\rm f.o.} \sim \left(\frac{m_{3/2}^4 M_P^3}{M_{\tilde{\gamma}}^2} \right)^{1/5} \sim 450 \text{ GeV} \left(\frac{m_{3/2}}{0.1 \text{ eV}} \right)^{4/5} \left(\frac{100 \text{ GeV}}{M_{\tilde{\gamma}}} \right)^{2/5}.$$

At which gravitino mass is $T_{\rm f.o.} \sim M_s \sim M_{\tilde{\gamma}}$?

Exercise 151. Estimate the largest possible gamma-ray flux for a 100 GeV gravitino pair-annihilating in the Galactic center.

What this means is that, not surprisingly, light gravitinos decouple at a very large temperature, much larger than their mass, and are thus hot relics. The resulting thermal relic abundance is then exactly what we estimated for a generic hot relic,

$$\Omega_{3/2}h^2 = \frac{m_{3/2}n_{3/2}(T_0)}{\rho_c}h^2 \simeq \frac{m_{3/2}}{\text{keV}}\left(\frac{100}{g_*(T_{\text{f.o.}})}\right). \tag{9.1}$$

Thus, $m_{3/2} \sim 100$ eV gravitinos have the "right" thermal relic abundance, but obviously do not work as a dark matter candidate — they are both hot and too light for a fermionic candidate from Fermi-blocking arguments (see Sec.1.9). However, this estimate has a flaw. As you calculated (or should have calculated) above, for $m_{3/2} \gtrsim 0.02$ eV, one has $T_{\text{f.o.}} \sim 100$ GeV. At those temperatures, s-particles (hopefully some of them) are in thermal equilibrium; at that epoch there exist processes involving *one* gravitino which are important to dump gravitinos in the primordial soup. The fastest such processes involve gauge superfields, i.e. gauge bosons V and their gauginos $\tilde{\lambda}$, for example

$$V + \tilde{\lambda} \leftrightarrow V + \tilde{G}; \qquad V + V \leftrightarrow \tilde{\lambda} + \tilde{G}.$$

The total cross section for these processes is something like [222]

$$\Sigma_{\text{tot}}^{(3/2)} \frac{\sum_{i=1,\ldots,3} c_i g_i^2 M_i^2}{24\pi(m_{3/2}M_P)^2},$$

where the c_i's are factors of order 1–30, and the i runs over the gauge couplings and gaugino soft supersymmetry breaking masses for the U(1), SU(2), and SU(3) gauge groups of the SM. Interestingly, the cross section does not depend on temperature, and is driven by gluino processes (the corresponding $c_i g_i^2$ being much larger than for binos and winos). Note that the cross section above goes like $(m_{3/2}M_P)^{-2} \sim F^{-2}$, with $F \sim M_s^2$ the supersymmetry breaking scale (squared), thus what actually gets produced are the *longitudinal* (Goldstino) modes: in the "global limit" $M_P \to \infty$ when one turns off gravity, those processes persist, since they originate from supersymmetry breaking!

Exercise 152. Show that the gravitino freeze-out temperature including single-gravitino processes is

$$T_{\rm f.o.} \simeq \frac{m_{3/2}^2 M_P \sqrt{g_{*S}^{\rm MSSM}}}{\alpha_s M_3^2} \sim 1\ {\rm TeV}\left(\frac{m_{3/2}}{1\ {\rm keV}}\right)^2 \left(\frac{1\ {\rm TeV}}{M_3}\right)^2,$$

where the latter assumes $g_{*S}^{\rm MSSM} = 228.75$.

The decoupling temperature you just computed illustrates that light gravitinos are in thermal equilibrium down to relatively low temperatures — compared to what e.g. we think the reheating temperature is. Eq. (9.1) then implies that gravitinos above a keV or so should have never thermalized, or they would be much too abundant and "overclose" the universe (in other words $\Omega_{3/2}h^2 \gg \Omega_{\rm DM}h^2$ for $m_{3/2} \gtrsim 1$ keV).

What is the fate of heavier gravitinos then? Things, as it turns out, start getting quite a bit more model dependent. Key processes are, once again, interactions with gauge superfields, as well as from the decay of heavier sparticles to gravitinos.

Exercise 153. Argue that the decay width of a sparticle of mass $m_{\tilde{s}}$ to a gravitino plus a SM particle (e.g. $\tilde{\lambda} \to \tilde{G} + V$ or $\tilde{f} \to \tilde{G} + f$) should go as (up to a phase-space factor and other multiplicative factors)

$$\Gamma \propto \frac{m_{\tilde{s}}^5}{(m_{3/2} M_P)^2}.$$

Assuming the inflaton ϕ does not decay directly into gravitinos, the system of differential equations one needs to solve to calculate the gravitino relic abundance is

$$\begin{cases} \dot{\rho}_\phi + 3H\rho_\phi = -\Gamma_\phi \rho_\phi, \\ \dot{\rho}_R + 4H\rho_R = +\Gamma_\phi \rho_\phi, \\ \dot{n}_{3/2} + 3H n_{3/2} = \gamma, \end{cases}$$

where Γ_ϕ is related to the reheating temperature $T_{\rm RH}$ via

$$\Gamma_\phi = H_R \equiv \frac{1}{M_P}\sqrt{\frac{8\pi}{3}\rho_R(T_{\rm RH})},$$

and $\gamma \propto T^6/M_P^2$ is a rate for gravitino production that includes both decay to gravitinos and $2 \leftrightarrow 2$ processes. To my knowledge, the most up-to-date calculation of γ is given in [217], and in units of $T^6/\bar{M}_P^2$ it is around 0.5 for a wide range of temperatures, for reasonable assumptions about sparticle masses. Solving the system of equations above one gets

$$\Omega_{3/2}h^2 \simeq 0.00167 \frac{m_{3/2}}{1 \text{ GeV}} \frac{T_{\rm RH}}{10^{10} \text{ GeV}} \frac{\gamma|_{T=T_{\rm RH}}}{T_{\rm RH}^6/\bar{M}_P^2}. \tag{9.2}$$

Exercise 154. Show that $\Omega_{3/2}h^2 \sim T_{\rm RH}$ (see e.g. the discussion in [223], Chapter 16.4).

In addition to the gravitino population described above, next-to-lightest supersymmetric particles (NLSP) decaying *after* they freeze out with a thermal relic abundance $\Omega_{\rm NLSP}h^2$ would dump extra gravitinos, with an abundance

$$\Omega_{3/2}h^2 = m_{3/2}\frac{\Omega_{\rm NLSP}h^2}{m_{\rm NLSP}}.$$

Such late decays might potentially upset predictions for the abundances of light nuclei from Big Bang Nucleosynthesis (BBN), and there are constraints on $\Omega_{\rm NLSP}h^2$ that depend on the nature of the NLSP and on its lifetime.

The bottom line of this discussion is that $m_{3/2} \ll 1$ keV are ruled out, while there are generic combinations of $m_{3/2} \gg 1$ keV and $T_{\rm RH}$ for which gravitinos are perfectly fine *cold* (or, if light enough, potentially *warm*) dark matter candidates.

Another possibility is that gravitinos are *not* the lightest supersymmetric particles (LSPs). The gravitino lifetime then depends on the particle nature of the LSP, but it will scale as

$$\tau_{\tilde{G}} \sim M_P^2/m_{3/2}^3 \sim 10^{14} \text{ sec} \left(\frac{1 \text{ GeV}}{m_{3/2}}\right)^3.$$

For heavy enough gravitinos, say $m_{3/2} \gg 100$ TeV, the decays happen before BBN, and are thus safe. One then simply has a population of non-thermally produced LSPs given by Eq. (9.2), scaled by the ratio of the gravitino-to-LSP mass. Lighter gravitinos are constrained by the effect of gravitino decays on BBN (or, if at very late times, on the cosmic microwave background (CMB) energy spectrum). As a result one gets bounds on $T_{\rm RH}$, which, at a given mass, fixes the gravitino abundance.

There are several other dark matter candidates that interact super-weakly both with themselves and with other SM particles — examples include Kaluza–Klein gravitons in universal extra dimension (UED), axinos in supersymmetry, "quintessinos", and many others. Depending on the particle physics model embedding, a thermal, freeze-in-like or non-thermal production mechanism can be in place for any one of those candidates. While there are really interesting pieces of physics that can be learned, alas, getting into details might be distracting for the purpose of this book and of this chapter in particular. So let us now move on from cosmology to experiments.

How can we possibly detect gravitinos or other super-weakly interacting dark matter particles? *Is there any hope?* Of course, pair-annihilation of gravitinos (i.e. some flavor of indirect detection), or direct detection, is hopeless.

Exercise 155. Estimate

(i) the gravitino-proton scattering cross section relevant for direct detection, and
(ii) the gravitino pair-annihilation cross section, relevant for indirect detection,

and convince yourself of said hopelessness.

However, some channels might exist. For example, the additional population of gravitinos produced by NLSP decays described above would be associated with the injection of energy in the early universe that could distort the CMB spectrum, or that could help synthesize rare light elements such as ^{6}Li and ^{7}Li [224]. If most of the gravitinos come from NLSP decays, then structure formation might be affected because the dark matter is produced at late times with large velocities. The two key ways in which structure formation is affected is by suppressing the linear power spectrum at small scales, and by reducing the phase-space density of halos, which could manifest itself as removing the central cusps of dark matter halos [225].

There are also other possible ways to detect gravitinos. An interesting possibility is to pair produce sparticles via reactions initiated by ultra high-energy cosmic rays (UHECR) neutrinos: consider for example the reaction

$$\nu + q \to \tilde{l} + \tilde{q},$$

where ν is an impinging ultra-high energy neutrino hitting a quark belonging to a nucleus in the atmosphere, and where $\tilde{l}, \tilde{q}$ are sleptons and squarks, the latter potentially promptly decaying into sleptons. The slepton decay width is

$$\Gamma(\tilde{l} \to l\tilde{G}) = \frac{1}{48\pi\bar{M}_P^2} \frac{m_{\tilde{l}}^5}{m_{3/2}^2} \left(1 - \frac{m_{3/2}^2}{m_{\tilde{l}}^2}\right)^4 .$$

This gives very long lifetimes, as you will show in the following exercise

Exercise 156. Show that for gravitino masses close to the slepton mass, $m_{\tilde{G}}/m_{\tilde{l}} \approx 1$, the slepton lifetime is

$$\tau(\tilde{l} \to l\tilde{G}) \simeq 4 \times 10^8 \text{ sec} \left(\frac{100 \text{ GeV}}{m_{\tilde{l}} - m_{\tilde{G}}}\right)^4 \left(\frac{m_{\tilde{G}}}{1 \text{ teV}}\right).$$

One would then see these massive charged particles propagate through neutrino telescopes such as IceCube, at some typically large transverse separations, leaving spectacular signals. The only question is: would the rate be large enough? You will answer this question next.

Exercise 157. Estimate the production rate for slepton pairs using a neutrino flux

$$\frac{d\phi_\nu}{dE} \sim 1\frac{10^{-8}}{E^2} \text{ GeV cm}^{-2} \text{ s}^{-1} \text{ sr}^{-1}$$

and a reasonable estimate for the cross section $\sigma(\nu + q \to \tilde{l} + \tilde{q})$. (if in need, see Ref. [226]).

Exercise 158. Estimate the slepton track separation $\delta R \simeq L\theta$ for a distance to production point $L \sim R_\odot$ and $\theta \simeq p_{\rm SUSY}^{\rm CM}/p_{\rm boost}$ the typical angle between the sleptons in the lab frame. Show that the ballpark is $\delta R \simeq$ few $\times$ (10–100) m.

A really interesting way to search for long-lived sleptons with colliders was put forth in Refs. [227, 228]: suppose you built giant tanks around Large Hadron Collider's (LHC's) collider detectors. If one can trap enough sleptons in those tanks, and if the sleptons decay on reasonably short timescales,[f] a signal from the tanks can be detected!

Exercise 159. Assume 1000 staus per year were produced at the LHC, and that the range for the produced staus in water is an average 10 m. Estimate the number of sleptons trapped in spherical-shell traps of inner radius 10 m, for trap volumes of 0.1, 1, and 10 kton. See [229] for details.

A final possibility is that gravitino dark matter not be absolutely stable, but rather decay via, for example, some R-parity violating operator. Gravitinos would be very long lived, from the suppression coming both from the usual powers of inverse Planck scale, and from the necessarily small (because of proton decay constraints) R parity-violating coefficients. Yet, potentially one could detect a continuum or line-like gamma-ray signal (depending on the decay mode) to be compared with bounds on the dark matter lifetime from Fermi-LAT observations which are just below $\tau \sim 10^{30}$ s [230].

9.3 Superheavy dark matter: WIMPzillas!

Particle pair-production in the presence of an intense, classical external field is a well-known quantum mechanical phenomenon: think about electron–positron pair production by strong electromagnetic fields. It was no one else than E. Schrödinger himself, back in 1939, to entertain the possibility of particle creation in the early universe by strong external *gravitational* fields [231], in a very evocatively titled article *"The proper vibrations of the expanding universe"*. Almost a half century ago this idea was put on firm grounds (see e.g. [232–235]). More recently, non-thermal super-heavy particle creation was reconsidered and acknowledged as a generic phenomenon in the context of inflationary cosmology [236], and given a rather catchy name (*"WIMPzillas!"*, [237]).

Instead of relying on thermal equilibrium, superheavy dark matter scenarios postulate two simple ingredients:

(i) the particle's lifetime is much longer than the age of the Universe,

---

[f] Here, the benchmark is one PhD thesis, or approximately 10^8 s.

(ii) the dark matter particle's interaction rate is weak enough that thermal equilibrium is *never* attained.

Interestingly, a generic prediction of superheavy dark matter models is to produce isocurvature perturbations, making this possibility seriously exciting after the tentative signal reported by the BICEP2 Collaboration [238] (which, however, is probably to be ascribed to polarized light emitted by Galactic dust [239]).

Let us think a little bit about condition (ii), i.e. that this superheavy wimpzilla monster X never be in thermal equilibrium. We know how to wrap our head around this in a quantitative way: we must demand that the particle scattering rate always be smaller than the expansion rate, i.e. $n_X \sigma v < H$ at all temperatures, with the meaning of the symbols I use which should be clear by now.[g] Call $T_* < M_X$ the temperature at which the wimpzillas were created, and assume that the universe was radiation dominated at $T = T_*$. We will (actually, you will!) show that, as long as the wimpzilla is heavy enough (to be quantified), and if the X was produced with a density low enough that today $\Omega_X h^2 < 1$, then the wimpzilla is *never* in thermal equilibrium.

Exercise 160. Start by showing (err, there is very little to show...) that

$$\Omega_X h^2 = \Omega_\gamma h^2 (T_*/T_0) m_X n_X(T_*)/\rho_\gamma(T_*),$$

where T_0 is the present temperature. Recast $\rho_\gamma(T_*)$ in terms of the Hubble rate at T_* and of T_*, and show that

$$\rho_\gamma(T_*) = (\Omega_X/\Omega_\gamma) T_0 M_P T_*/M_X.$$

Now invoke unitarity and use $\sigma v < M_X^{-2}$ to show that X is never in thermal equilibrium as long as

$$\left(\frac{200\ \text{TeV}}{M_X}\right)^2 \left(\frac{T_*}{M_X}\right) < 1.$$

The exercise above shows that if a mechanism exists where the X is produced with a small enough density that today $\Omega_X h^2 < 1$, then condition (i) is *automatic*! Now, let us focus on that critical moment when wimpzillas are created. The most likely point in time for that to happen is just after

[g] If it is not, you win a free pass to Chapter 2!

inflation. The task is thus to estimate the energy density in wimpzillas when the universe is reheated after inflation.

Assuming instantaneous reheat, i.e. that the inflaton field's energy density is converted instantaneously into radiation when the decay rate of the inflaton $\Gamma_\phi = H$, it is an instructive exercise to calculate $T_{\rm RH}$ as a function of Γ_ϕ.

Exercise 161. Use Friedmann's equation at $T_{\rm RH}$, the condition $H(T_{\rm RH}) = \Gamma_\phi$, and the condition that at $T_{\rm RH}$ the energy density $\rho_\phi = \rho_R$ to show that

$$T_{\rm RH} = \left(\frac{90}{8\pi^3 g_*}\right)^{1/4} \sqrt{\Gamma_\phi M_P} \simeq 0.2 \left(\frac{200}{g_*}\right)^{1/4} \sqrt{\Gamma_\phi M_P}.$$

There are constraints on $T_{\rm RH}$. For example, Ref. [246] finds that

$$T_{\rm RH} \gtrsim 10^7 \left(\frac{0.2}{r}\right) \text{ GeV},$$

with r the tensor to scalar ratio; gravitino overproduction (see Sec. 9.2) would imply $T_{\rm RH} \lesssim 10^9$–10^{10} GeV [237]. The calculation of the density of X particles produced via gravitational interactions[h] at the end of inflation is model-dependent, and we will not reproduce it here at all. Wise people [240, 241] provide us with the following useful and, as it turns out, rather generic expression:

$$\rho_X(T_*) \sim 10^{-3} M_X^4 \left(\frac{M_X}{M_\phi}\right)^{-3/2} e^{-2M_X/M_\phi},$$

which in turn, as you will show in the following exercise, gives

$$\Omega_X h^2 \simeq 10^{-3} \Omega_R \frac{8\pi}{3} \left(\frac{T_{\rm RH}}{T_0}\right) \left(\frac{M_\phi}{M_P}\right)^2 \left(\frac{M_X}{M_\phi}\right)^{5/2} e^{-2M_X/M_\phi}, \tag{9.3}$$

with Ω_R the radiation energy density to critical energy density.

[h]More precisely, by the interaction of a classical, time-dependent gravitational field with a quantum vacuum; clearly this depends both on the choice of vacuum, hence on the quantum theory, and on the structure of the time-dependence, i.e. on the cosmology.

Exercise 162. With your newly acquired knowledge, demonstrate Eq. (9.3).

The annihilation rate for wimpzillas is tiny: any annihilation signal would be puny compared to, say, a vanilla WIMP for these massive, non-thermally produced particles (using the non-thermalization value for the mass we estimated above $M_X \gtrsim 200$ TeV).

Exercise 163. Consider a wimpzilla X with $\sigma v \sim M_X^{-2}$ and a WIMP χ with $\sigma v \sim G_F^2 m_\chi^2$. Show that the ratio of the pair-annihilation rate per unit volume, $n^2\sigma v$, for X over χ goes like $G_F^2 M_X^4 \sim 10^{11} \left(\frac{M_X}{200 \text{ TeV}}\right)^4$.

Nevertheless, these massive particles might decay and produce "interesting" debris that could be detected. For sensible combinations of $T_{\rm RH}$, M_ϕ, M_X and the X lifetime $1/\Gamma_X$ one can, at once, obtain a good relic density, and an interesting flux for detection with UHECR detectors like Auger. Let us make a couple estimates.

The flux of cosmic rays from the decay of a super-massive particle of mass M_X is given by the usual integral over the number density of particles along the line of sight, times the decay rate Γ_X, times a fragmentation function $dN(E, M_X)/dE$ that can be extrapolated via standard quantum chromodynamics (QCD) techniques: for example, in the modified leading-logarithmic approximation, Ref. [242] finds

$$\frac{dN(x)}{dx} = \frac{15}{16} x^{-3/2}(1-x)^2 \sim x^{-3/2},$$

with $x = E/E_{\rm jet}$ and, for a decaying particle $E_{\rm jet} = M_X/2$ (for annihilation, of course, $E_{\rm jet} = M_X$). This allows you to calculate how the flux of junk produced in a wimpzilla decay scales with the wimpzilla particle mass (if in doubt see [243]):

Exercise 164. Show that the flux of decay debris from a wimpzilla of mass M_X scales as

$$\phi \sim \Gamma_X M_X^{-1/2},$$

and that the annihilation debris would scale, instead, as

$$\phi \sim \langle \sigma_A v \rangle_X M_X^{-3/2}.$$

(The only thing you really need to show in this exercise is that the fragmentation function gives a factor $M_X^{1/2}$ in both cases).

Given this, you have all of the ingredients for the estimate we are after:

Exercise 165. Assuming that you wanted to explain a flux

$$E^3 \phi(E) \sim 10^{24}\ \mathrm{eV^2 m^{-2} s^{-1} sr^{-1}},$$

with the decay of Galactic wimpzillas of mass $M_{X_1} = 10^{12}$ GeV and $M_{X_2} = 10^{14}$ GeV, show that the required lifetime is $\Gamma_{X_1} \sim 10^{20}$ yr and $\Gamma_{X_1} \sim 10^{21}$ yr, respectively.

There are other interesting super-heavy dark matter candidates. I especially like a class of candidates called "strangelets". These are literally nuggets of quarks that could form in a first-order phase transition (as first envisioned by Witten, [244]), and be stable. When we say nuggets we mean macroscopic clumps of quark matter of radius between 10 mm and 10 cm, weighing somewhere in the range 10^9–10^{18} g (that is a lot).

Exercise 166. Show that the rate of collision of strangelets with the Earth ranges between one per year (for small ones of 10 mm radius) and one per billion year (for the biggies of 10 cm radius).

A strangelet would fly through the Earth poking a hole the size of its geometric size and displacing everything on its path (stuff the Earth is made of is many orders of magnitude less dense than the strangelet). But such events can easily be rare enough for strangelets to be viable dark matter candidates.

There are two reasons why I like strangelets. The first one is that perhaps it is possible to attach to a strangelet dark matter framework successful baryogenesis; and the second one is that astrophysical signals from strangelets are quite interesting and unique. Let me tell you a little bit about both items.

Baryogenesis can arise from the effect of the QCD θ term (see Chapter 7) in biasing anti-quark nuggets over quark nuggets.[i] Quantitatively, if the ratio of nuggets to antinuggets is 2:3 at the end of nugget formation, baryogenesis is explained without needing any net baryon number excess (the antibaryon excess is captured in the antinuggets), and dark matter is explained by the nuggets plus anitnuggets [245].

Exercise 167. Using a geometric cross section, argue that nuggets–antinuggets pair-annihilation in the late universe is a very rare, and undetectable process.

The typical baryon number content of these nuggets is $|B| \gtrsim 10^{25}$, so the effective interaction rate (σ/M) is always too small to be of any relevance for astrophysics or cosmology.

Exercise 168. Estimate σ/M for quark nuggets of $|B| \gtrsim 10^{25}$.

However, cosmic rays, free electrons and interstellar gas can all hit strangelets producing detectable radiation, from radio up to gamma-ray frequencies. The morphology of the emission is pretty unique, as it traces the *product* of the ordinary matter times dark matter density (some of the astrophysical excesses that might be associated with nuggets dark matter emission do not have that morphology, though [246]), and as you will show below the nuggets flux is not dissimilar to the flux of cosmic rays near the Greisen–Zatsepin–Kuzmin (GZK) limit. Let us first make sure we understand what "GZK limit" actually means, and then play with these nuggets' numbers.

[i] Interestingly, this could be tested at the Relativistic Heavy Ion Collider (RHIC) and at the LHC! See [245].

Exercise 169. The GZK limit, or cutoff, is the energy at which cosmic-ray protons from distant, extra-galactic sources are expected to be suppressed by interactions with the CMB photons. In practice, this corresponds to the reaction

$$\gamma_{\rm CMB} + p \to p + \pi^0 \text{ (or } n + \pi^+).$$

Calculate the threshold incident cosmic-ray proton energy for the process to happen. That energy roughly corresponds to the GZK limit.

Exercise 170. Show that the flux of quark nuggets scales as

$$\frac{dN}{dA\,dt} = nv \approx \left(\frac{10^{25}}{B}\right) \text{ km}^{-2}\text{yr}^{-1}.$$

Exercise 171. Assuming a Galactic gas density of 10^3 cm^{-3} in a certain direction and a dark matter density of 10 GeV cm^{-3}, estimate the flux of radiation from gas-nuggets collisions.

Another interesting and theoretically very well motivated class of supermassive dark matter candidates is that of Q-balls. I will not review them here, but rather point the interested Reader to a great review, Ref. [247]. Again, one of the phenomenological motivations for Q-balls is that one can explain baryogenesis and dark matter in one move, and connect that explanation to inflation, for extra gravy and extra kudos!

9.4 Self-interacting dark matter (SIDM)

When is dark matter self-interaction relevant to the level of affecting the macroscopic properties of dark matter halos? A simple rule of thumb is to have at least one interaction per particle over the age of the universe. What this means, quantitatively, for a halo with typical velocity v and typical density ρ, is that

$$\Gamma/H_0 = \frac{(\sigma_{\chi\chi}/m_\chi)\rho v}{H_0} \sim 1,$$

which means having

$$\left(\frac{\sigma_{\chi\chi}}{m_\chi}\right) \sim 1.3\frac{\text{cm}^2}{\text{g}}\left(\frac{1\ \text{GeV/cm}^3}{\rho}\right)\left(\frac{10\ \text{km/s}}{v}\right).$$

Exercise 172. Cross check the estimate above and reconstruct the $1.3\,\text{cm}^2/\text{g}$.

Different halos have different combinations of $\rho \cdot v$, and will thus be sensitive to different ratios of $\sigma_{\chi\chi}/m_\chi$. In some sense, dark matter halos of different sizes provide different "collider" probes for dark matter self-interactions. My friend and colleague Hai-Bo Yu has a very suggestive way of drawing a parallel between dark matter halos of different sizes and colliders at different center-of-mass energies, whereby dwarf galaxies ($v \sim 30$ km/s) are the analog of "B-factories", Milky-Way-size galaxies ($v \sim 200$ km/s) of "LEP" and clusters of galaxies of the "LHC" ($v \sim 1000$ km/s).

Observationally, whether or not dark matter self-interactions are motivated or even needed is at the moment rather unclear (see e.g. the recent review [248]). However, there are a few interesting tentative indications, from e.g. observations of cored dark matter profiles in small systems (dwarf galaxies, low-surface brightness spiral galaxies) and even in certain well-resolved galaxy clusters (see e.g. [249]), and from the so-called too-big-to-fail problems (N-body simulations for cold dark matter indicate an excess of large sub-halos that have no visible counterpart in observed Milky-Way satellite galaxies). Simulations that employ $\sigma_{\chi\chi}/m_\chi \sim 0.5$–$50$ cm^2/g show suppression both in the central dark matter density of halos, and affect the statistics of sub-halos in a way that facilitates agreement with observations.

The needed self-interaction cross sections are very large. They are basically nuclear-scale cross sections (just remember that 1 cm^2/g $\sim$ 1.8 barn/GeV!). As we indicated in Sec. 1.9, observations on clusters scales constrain $\sigma_{\chi\chi}/m_\chi \lesssim 1$ cm^2/g for $v \sim 3000$ km/s (or for constant $\sigma(v)$). It is interesting to think about models where such large self-interaction cross section are possible, and whether or not, in addition, it is possible to have a velocity dependence such that one can evade constraints from clusters (large v) and have large effects on small scales (low v).

A simple example of strong-interaction-strength cross section is to just postulate a strongly interacting *hidden* sector, the simplest possibility being an SU(N) gauge symmetry with no matter content at all (See Ref. [250]).

What happens in this theory is quite simple: at the confinement scale Λ when the theory becomes strongly coupled, the hidden gluons form bound states, "glueballs". At low energies $\ll \Lambda$ the model is described by an effective field theory of the lowest-mass such states, with typical masses $m \sim \Lambda$. Naive dimensional analysis implies that the self-interaction cross sections for the stable lowest-mass glueball gb, i.e.

$$\sigma(gb + gb \leftrightarrow gb + gb) \sim \frac{4\pi}{\Lambda^2},$$

be essentially geometric, and velocity-independent.

Exercise 173. Show that

$$\left(\frac{\sigma}{m}\right)_{gb} \simeq 2.7 \frac{\text{cm}^2}{\text{g}} \left(\frac{0.1\text{ GeV}}{\Lambda}\right)^3.$$

The relic density of glueballs is easily computed:

Exercise 174. After confinement, the massless hidden gluons are bound into glueballs of mass $m \sim \Lambda$, which naturally become non-relativistic for hidden-sector temperatures $T_h \ll \Lambda$. Assuming the number density of glueballs is the same as that of gluons before the phase transition, and assuming there are no number-changing processes (e.g. $3 \to 2$ scatterings), show that the relic density of glueballs is

$$\rho_{gb} \sim s_0 \Lambda (2(N^2 - 1)) \left(\frac{T_h}{T}|_{T\sim\Lambda}\right)^3.$$

Unfortunately, in this model there is no "connection" between the hidden and the visible sector. Ref. [250] discusses interesting possible extensions to a strongly interacting hidden sector including matter fermions and connector fields. Worth reading!

Another possibility is that there be a light particle ϕ mediating scattering between dark matter particles X. The relevant parameters of the theory are then the mediator mass m_ϕ, the dark matter mass m_X, and the coupling α_X. This model would imply two things:

1. A self-interaction cross section, in the perturbative limit and at low velocity ($m_X v \ll m_\phi$), of the order of

$$\sigma \sim 5 \times 10^{-23}\ \text{cm}^2 \left(\frac{\alpha_X}{0.01}\right)^2 \left(\frac{m_X}{10\ \text{GeV}}\right)^2 \left(\frac{10\ \text{MeV}}{m_\phi}\right)^4;$$

2. In the opposite, high-velocity limit ($m_X v \gg m_\phi$), one finds instead

$$\sigma \sim \frac{\alpha_X^2}{m_X^2 v^4},$$

producing a (strongly) velocity-dependent cross section, and a Yukawa-like interaction potential

$$V(r) = \pm \frac{\alpha_X}{r} e^{-m_\phi r}.$$

Exercise 175. Convince yourself that the cross section estimates above do indeed work.

One catch to this story is that if the mediator is stable it could easily dominate the energy density of the universe. If one postulates that the mediator is a scalar, and that it couples to SM fermions through a vertex with a dimensionless coupling ϵ, for sufficiently large ϵ, the ϕ will decay before upsetting BBN:

Exercise 176. Show that the requirement that the lifetime of ϕ be shorter than $\sim$1 second (the typical BBN timescale) implies

$$\epsilon \gtrsim 10^{-10} \sqrt{10\ \text{MeV}/m_\phi}.$$

In this simple model, the dark matter scatters through ϕ-exchange off of nuclei:

Exercise 177. Show that the direct detection cross section

$$\frac{d\sigma}{dq^2} = \frac{4\pi \alpha_{\text{EM}} \alpha_X \epsilon^2 Z^2}{(q^2 + m_\phi^2)^2 v^2}.$$

It turns out that in this setup, if $m_X \gtrsim 10$ GeV, combinations of parameters giving an interesting self-interaction cross section produce a direct detection cross section excluded by current experiments. Light ($m_X \ll 10$ GeV) dark matter candidates, instead, do work.

Exercise 178. Run the numbers to demonstrate the statement above. In doubt or for help see [251].

A strong velocity dependence induces peculiar nuclear recoil spectra as discussed in Sec. 4.4, and can thus be differentiated from standard WIMP "contact interaction" type cross section signals. If (boosted) ϕ's from $X + X \to \phi + \phi$ decay into electrons, one would also expect an inverse compton signal at gamma-ray energies.

Exercise 179. Calculate the typical gamma-ray energy for electrons produced in the annihilation of a $m_X \sim 100$ GeV particle to $\phi \to e + e^-$ up-scattering interstellar radiation photons of energy $E_{\rm ISRF} \sim 1$ eV.

One last interesting, simple SIDM model gives us the opportunity to refresh our memory about spontaneous symmetry breaking[j]: consider a complex scalar field Φ with a potential

$$V(\Phi) = -\mu^2|\Phi|^2 + \frac{\lambda}{4}|\Phi|^4.$$

After spontaneous symmetry breaking, call $v = \langle\Phi\rangle = \mu\sqrt{2/\lambda}$, and define the real scalar components of Φ as

$$\Phi = v + \frac{s + ia}{\sqrt{2}}.$$

Exercise 180. Show that the potential with the field redefinition above, up to unphysical constants, reads

$$V(s,a) = \frac{m_s^2}{2}s^2 + \frac{\sqrt{\lambda}}{2\sqrt{2}}m_s s^3 + \frac{\sqrt{\lambda}}{2\sqrt{2}}m_s a^2 s + \frac{\lambda}{16}s^4 + \frac{\lambda}{16}a^4 + \frac{\lambda}{8}a^2 s^2,$$

with $m_s = \sqrt{2}\mu$.

[j] See Ref. [252] for more details on this model.

The massless pseudoscalar a is assumed to acquire a mass $m_a \ll m_s$ from non-perturbative effects which break the model's U(1) symmetry down to Z_N. Now let us estimate the self-interaction cross section $a + a \leftrightarrow a + a$:

Exercise 181. This is a lengthy, but worthwhile, exercise to follow through...

(i) Write down the four Feynman diagrams for a self-interaction;
(ii) Guess the scaling of the cross section with λ, m_a, and m_s, and show that the cross section $\sigma(a + a \leftrightarrow a + a) \to 0$ for $m_a \to 0$; interpret the origin of the cancelation of the $1/m_a^2$ divergence in the small m_a limit with a SM analog (see Ref. [258] if in need!);
(iii) Do the full calculation and show that

$$\sigma(a + a \leftrightarrow a + a) = \frac{\lambda^2 m_a^2}{32\pi m_s^4} \left(1 - 4\frac{m_a^2}{m_s^2}\right)^{-2};$$

(iv) Set $m_s = 1$ MeV and $\lambda \simeq 0.5$. Calculate m_a such that

$$\frac{\sigma(a + a \leftrightarrow a + a)}{m_a} \sim 1 \, \frac{\text{cm}^2}{\text{g}}.$$

9.5 Asymmetric dark matter

We have no direct indication of what caused the observed excess of baryons over antibaryons. However, this open problem could be turned into an opportunity: perhaps the baryon asymmetry is related to the mechanism that produced dark matter with the cosmologically observed abundance. The simplest possible idea is that the dark matter is the *asymmetric* leftover of an imbalance between dark matter and anti-dark matter, and that such an asymmetry is related to that in the baryonic sector. The earliest incarnation of this idea came with the fancy name *"Technocosmology"* [253].[k] Once again, this old idea was recently rediscovered and made popular again [254], leading also to several interesting, and *new* developments (see Ref. [255] for a review).

---

[k]The dark matter candidate was the lightest "technibaryon", belonging to a sector that shared a quantum number with ordinary baryons [253].

Asymmetric dark matter models can be broadly classified according to whether or not there is a net global (generalized) baryon asymmetry. Let us be more definite: suppose there exists a global additive quantum number, a "dark baryon number" B_D, preserved in the low-energy interactions of the dark matter sector, in addition to the usual baryon number of the visible sector (let us indicate the latter with B_V). The dark matter particle shall be the *lightest particle* charged under B_D. Now consider the following two linear combinations:

$$B_{\text{conserved}} = B_V - B_D;$$
$$B_{\text{broken}} = B_V + B_D.$$

A first possibility is that one linear combination, $B_{\text{conserved}}$, is effectively unbroken at all energies, while the second one is broken at some high temperature. At those temperatures, one also assumes the occurrence of the other required Sakharov conditions to create a net baryon asymmetry in the universe [256]: interactions must also violate C and CP, and occur out of thermal equilibrium. If this is the case, some net charge ΔB_{broken} is produced, and since $B_{\text{conserved}}$ is, ahem, conserved, it will distribute the asymmetry so that there are equal amounts in the visible and dark sector, $\Delta B_V = \Delta B_D = \Delta B_{\text{broken}}/2$. This is great: at low energy the B_{broken} interactions decouple, B_D and B_V are independent symmetries of the low-energy interactions, and the net charged ΔB_V and ΔB_D trickle down to the lightest stable particles in each sector. This scenario has been implemented for a variety of generalized-baryon-number-generating mechanisms, and for a variety of dark sectors [255].

A second possibility is that the universe is, instead, *not* generalized-baryon-number symmetric: rather, some asymmetry is generated either in B_V or in B_D, or even in both sectors separately (with the usual song and dance of C and CP breaking and out of equilibrium conditions), and at some intermediate energies there exist processes that only respect some linear combination of B_V and B_D, and thus redistribute the asymmetry(-ies), as long as the processes keep the particles in the two sectors in chemical equilibrium. These intermediate interactions then freeze out (i.e. they go out of chemical equilibrium) leaving a net global charge in the two sectors, which is left alone by low-energy $B_{D,V}$-conserving interactions. In this case, the relation between the visible- and dark-sector asymmetries depends on the details of the chemical decoupling, with important phenomenological consequences, which we briefly discuss below.

In addition to the symmetry structure above, to be relevant to the bottom-line tally, asymmetric dark matter should feature some mechanism that gets rid of the *symmetric* component. In general this depends on how efficient the dark matter–anti-dark-matter processes are, and thus on how large the pair-annihilation cross section $\sigma_{\bar{\chi}\chi}$ is.

Exercise 182. Show that the asymptotic, zero-temperature ratio for the anti-dark-matter to dark matter number density

$$r_\infty \equiv n_{\bar{\chi}}(T \to 0)/n_\chi(T \to 0),$$

obeys the equation

$$r_\infty \simeq \exp\left[-2\left(\frac{\sigma_{\bar{\chi}\chi}}{\sigma_{\rm WIMP}}\right)\left(\frac{1-r_\infty}{1+r_\infty}\right)\right] \xrightarrow{r_\infty \ll 1} \exp\left(-2\sigma_{\bar{\chi}\chi}/\sigma_{\rm WIMP}\right),$$

where $\sigma_{\rm WIMP}$ is the typical cross section needed to produce a thermal relic with the right leftover density in the symmetric case, $\sigma_{\rm WIMP}v \simeq 3 \times 10^{-26}\ {\rm cm}^3/{\rm s}$. If in doubt, see Ref. [257].

Thanks to the exponential dependence you found in Exercise 182, in practice one just needs a slightly larger-than-usual $\sigma_{\bar{\chi}\chi}/\sigma_{\rm WIMP} \gtrsim 1.4$ to have $r_\infty \lesssim 0.1$. Enforcing a relatively large pair-annihilation cross section can have important implications for phenomenology: the dark sector might include a light mediator responsible for the large annihilation rates, which would result, potentially, in dark radiation and thus in an additional component to the relativistic energy density (if stable), and/or produce detectable effects on small-scale structure via self-interactions (see Sec. 9.4).

What is the mass of asymmetric dark matter candidates?

Exercise 183. Show that the mass of asymmetric dark matter candidates is

$$m_{\rm DM} = m_p \frac{\Omega_{\rm DM}}{\Omega_b} \frac{\eta(B_V)}{\eta(B_D)/q_{\rm DM}} \frac{1-r_\infty}{1+r_\infty},$$

where m_p is the proton mass, η is the charge-to-entropy ratio, and $q_{\rm DM}$ is the dark-baryonic charge of the dark matter (see Ref. [258] if in need of help).

As you just showed in Exercise 183, if $r_\infty \to 0$ and if the universe is baryonic-symmetric, the dark matter mass is in the range[1]

$$m_{\rm DM} = q_{\rm DM} \times (1.6 - 5)\ \text{GeV}.$$

However, if instead we are in the second symmetry structure outlined above, the non-baryonic-symmetric case, then the general prediction of an $\mathcal{O}$(GeV) scale dark matter mass breaks down: in fact, unless chemical decoupling happens when particles in both the visible and dark sector are relativistic, the number density of the dark matter can be non-relativistic, and thus Boltzmann suppressed,

$$\eta(B_D)/\eta(B_V) \sim \exp(-m_{\rm DM}/T_D).$$

Note that in general the dark-matter temperature $T_D \neq T$, where T is the good old photon temperature. If this possibility is realized, the dark matter mass can easily fall in the TeV range.

Asymmetric dark matter can have rather distinctive, albeit model-dependent phenomenological features. These include peculiar recoil spectra in direct detection experiments for models with a light mediator, but such a prediction is by no means unique to an asymmetric dark matter generation mechanism. Another interesting possibility is that the interactions responsible for the generation of asymmetries in the dark and visible sector could mediate asymmetric dark matter decays into matter and antimatter [259]. This might produce observable and rather unique features in cosmic-ray spectra (for example, one would never expect the positron fraction to exceed 0.5 neither in standard cosmic-ray models nor in "symmetric" dark matter models, but this is a possibility for asymmetric dark matter!). Finally, asymmetric dark matter can be captured in stars, affecting things like thermal conductivity, the sound speed, and stellar oscillation modes, or even produce a detectable flux of high-energy neutrinos from the Sun.

Exercise 184. Using what we calculated for WIMP capture from the Sun, estimate the scale of the nucleon-dark matter cross section $\sigma_{n\chi}$ necessary to have significant accumulation of (asymmetric) dark matter in stars. Compare what you find with, e.g. the results of Ref. [260].

---

[1]This range includes a potential suppression factor $a_s = \eta(B_V)/\eta(B_D) \simeq 0.35$ if the generation of the asymmetries happened before the electroweak phase transition (if it happened afterwords $a_s \simeq 1$).

It is interesting to ask the question of whether any "traditional" indirect dark matter detection signal from pair-annihilation could exist if dark matter originates from an asymmetry. If no anti-dark matter is generated (and if primordial pair-annihilation got rid of all of it), there are clearly no indirect signals whatsoever; however, if dark matter and anti-dark matter oscillate into each other (for example because of a $\Delta N_\chi = 2$ operator, such as a mass term, say $m_M \chi\chi$) then oscillations can populate the anti-dark matter content of the universe and residual annihilation occur [261]. Unless a symmetry in the theory explicitly prohibits such operators, oscillations generically occur, since after all we only need

$$\tau_{\rm universe} \sim 10^{17}\ {\rm s} \lesssim 10^{17}\ {\rm s}\left(\frac{10^{-41}\ {\rm GeV}}{m_M}\right), \quad \text{i.e. } m_M \gtrsim 10^{-41}\ {\rm GeV}.$$

Exercise 185. Explain the mass scale in the equation above.

9.6 What does "minimality" mean, and how does it look like?

The notion of minimality is an intrinsically tricky one.[m] How is the counting done when it comes to defining what is the most minimal extension to the SM of particle physics? (In other words, what, exactly, is to be minimized?) Not obvious. One possibility is to count by the number of additional *new physical degrees of freedom* — a real scalar field is, accordingly, "more minimal" than, for instance, a complex scalar field; another possibility is to count by the number of additional *new parameters* in the theory: a scalar field will come with a scalar potential, which will contain a certain number of new parameters, while, for example, an additional fermionic $SU(2)_L$ multiplet would only come with its gauge quantum number and a mass.

How about neutrinos?

The most minimal possibility of all, however, is to postulate that there be actually no "new" ingredients at all, and that the dark matter is just a heavier copy of the (SM) neutrino.[n] After all, in the wise words of Lars

[m]Almost as tricky as that of "naturalness".

Bergstrom [262], neutrinos *"have the undisputed virtue of being known to exist"*. As we discussed, SM neutrinos cannot make up the bulk of the dark matter, for a variety of reasons, but chiefly because they would produce the wrong structure formation history, and because the lower limit to the neutrino mass from the Tremaine–Gunn limit (based, remember, on phase-space density arguments) would make neutrinos overclose the universe.

Exercise 186. Prove the statement above reminding yourself of the Tremaine–Gunn limit, of the relic density of neutrinos as a function of their mass, and of the fact that you should not mess up with the thermal history of the universe below BBN.

Now, a neutrino which is heavy, but lighter than half the mass of the Z is definitely ruled out by the Z decay width, as you will shortly convince yourself of.

Exercise 187. Estimate the Z decay width for decay into a fourth generation SM-like neutrino with $m_\nu < m_Z/2$, and compare with the limits on the invisible decay width of the Z, $\Gamma(\text{invisible}) = 499.0 \pm 1.5\,\text{MeV}$ and conclude that a fourth generation neutrino is excluded.

What about a fourth generation, heavier SM-like neutrino with $m_\nu > m_Z/2$? In this case, the trouble comes from direct detection: a Dirac neutrino has a spin-independent cross section off of nucleons of the order of $\sigma_{\nu n} \sim G_F^2 m_p^2 \sim 10^{-38}\ \text{cm}^2$, as we have seen in Chapter 4, which is about seven orders of magnitude too big given current limits.

Exercise 188. If, as we argued and showed in Chapter 2 (see Fig. 2.3) $\Omega_\nu h^2 \sim 0.6(m_\nu/1\,\text{TeV})^2$, and if the current limits in the $m_\nu \gg m_Z/2$ region are approximately $\sigma_{\nu n} \lesssim 10^{-45}\ \text{cm}^2(m_\nu/100\ \text{GeV})$, find how heavy m_ν should be and how overabundant it would then be from thermal production.

[n]Of course, I am assuming there exists a mechanism producing masses and mixing for the SM neutrinos, as well as for the additional, heavier neutrino.

The exercise you just (hopefully) carried out illustrated that direct detection limits force a fourth generation neutrino to a very, very large mass, where it would badly overclose the universe in a standard cosmology. Is there any way out?

One possibility is to split the Dirac neutrino eigenstates with a small Majorana mass; or to postulate an additional new Z' and have the heavy neutrino be sterile under $SU(2)_L$ and only charged under the Z', which would then mix with the Z after electroweak symmetry breaking [263]; there are countless other possibilities, including for the SUSY partners of neutrinos etc. But then, alas, *Adieu Minimalisme!*

The real, singlet scalar

If we go with the notion of minimality as in minimal number of *new degrees of freedom*, a first possibility is to add one real scalar field S with no gauge interactions (S thus stands for "singlet") to the SM. Definitely cannot go smaller than that in terms of field degrees of freedom! However, even only considering renormalizable interactions and potential terms, one ends up with at least three physical free parameters, in a theory where the real singlet is a dark matter candidate, as we will shortly see. Incidentally, this setup is a generic, and minimal, realization of the so-called *"Higgs portal"* scenario — the dark matter interacts through the gauge-singlet operator $H^\dagger H$, and it is a good and simple such example. Let us work through it.

First, to ensure stability, we need to postulate an unbroken Z_2 symmetry under which $S \to -S$ and every other SM field does not change. This limits the renormalizable interactions with the SM to only one possible term. The Lagrangian of the model then reads

$$\mathcal{L} = \mathcal{L}_{\rm SM} + \frac{(\partial_\mu S)^2}{2} - \frac{b_2}{2}S^2 - \frac{a_2}{2}H^\dagger H S^2 - \frac{b_4}{4}S^4.$$

The Z_2 symmetry actually does not by itself guarantee the stability of the real scalar: should the vacuum expectation value $\langle S\rangle \neq 0$ after electroweak symmetry breaking, clearly the singlet scalar could decay through the quartic coupling (a_2) interaction if $S \to s + \langle S\rangle$.

Exercise 189.

(1) Which constraints on a_2 and b_4 can you derive from the requirement that the potential be bound from below?
(2) Which constraints on the model parameters can you derive from the requirement that, at tree level, $\langle S \rangle = 0$?

While the exercise above illustrates that the quartic self-interaction coupling b_4 is important to ensure a viable (bounded-from-below) potential, the b_4 parameter is largely non-influential for the phenomenology of the real scalar singlet s.

Exercise 190. Calculate the singlet mass m_s at tree level.

The problem above shows that we can adopt, as free parameters in the theory with some impact on phenomenology, the pair (m_s, a_2). The singlet annihilates through s-channel SM Higgs exchange, and scatters with SM fermions also via Higgs exchange. It is relatively simple to calculate the thermal relic abundance of the singlet, and its spin-independent scattering cross section off of nucleons. You are asked to calculate the latter in the following exercise.

Exercise 191. Show that the scalar interaction between s and nucleons is

$$\sigma_{\rm SI} = \frac{a_2^2 m_N^4 f^2}{\pi m_s^2 m_h^4},$$

where f is the nucleon N matrix element and $m_h \simeq 125$ GeV is the SM Higgs mass.

Another couple of simple calculations I invite you to sketch are the pair-annihilation cross section into two photons (relevant for indirect detection, see [264]) and the contribution to the invisible decay width of the SM Higgs (relevant if $m_s < m_h/2$, see [265]).

Exercise 192. Express the pair-annihilation cross section for $ss \to \gamma\gamma$ in terms of the decay width of the SM Higgs to two photons.

Exercise 193. Calculate the decay width for $h \to ss$ and compare with limits on $\Gamma_h^{\rm inv}$ (see Sec. 6.5).

In Ref. [264] it was shown that there is only one, small region of parameter space where s is a good thermal relic — close to the resonance at $m_s \sim m_h/2$ with $10^{-3} \lesssim |a_2| \lesssim 10^{-2}$, and a_2 is small enough to evade constraints from direct detection (see e.g. [266]). In the low- and high-mass regions direct detection constraints rule out a thermal relic, and the s's relic density must be smaller than the observed dark matter density. In the resonant region, $ss \to \gamma\gamma$ offers good prospects for detection with Fermi, and is not ruled out by current data [267].

The "original" minimal dark matter

Minimality in the direction of the fewest possible *additional new parameters* is a different story. Think about an $SU(2)_L$ multiplet, which interacts through SM gauge interactions only via its covariant derivative. In this case, the one and only parameter of the theory is the mass of the multiplet! This is indeed minimal in this other direction — one, real number is all you need to fully define the theory, in addition, of course, to a few discrete quantum numbers. It is then a rather interesting exercise, that some of my Italian friends and collaborators, and others, have carried out over the last few years, to define a "minimal" WIMP dark matter scenario by outlining a set of rules that define a "*minimal*" $SU(2)_L$ WIMP, and by working through the resulting phenomenology. The rules, roughly, can be summarized as follows:

1. The MDM particle is charged under the SM weak interactions' $SU(2)_L$, i.e. it belongs to an $SU(2)_L$ multiplet;
2. The MDM multiplet has electric charge zero;
3. The MDM is stable;
4. The MDM has the correct thermal relic density;
5. There is no additional new physics up to the Planck scale, and there should thus be no Landau poles up to the Planck scale.

Additionally, MDM is either a scalar or a fermion (although in principle higher spins should also be considered). The first item, and the latter assumption, force us to write down the Lagrangian of the theory (in addition, of course, to the SM Lagrangian) as:

$$\mathcal{L} = \bar{\chi}(i\not{D} + M)\chi \qquad \text{fermion,}$$
$$\mathcal{L} = |D_\mu \chi|^2 - M^2|\chi|^2 \quad \text{scalar.}$$

The D's of course are covariant derivatives that induce $SU(2)_L$ and $U(1)_Y$ gauge interactions with SM field. The theory, as such, is completely defined by one single parameter: M (although of course there might be additional terms in the scalar potential). Also, one must define the quantum numbers of the new particle under $SU(2)_L$ and $U(1)_Y$. Let us start with $SU(2)_L$: this is just as good old spin, so we will just need to define one quantum number, the dimension n of the multiplet ($n = 2S + 1$ in the familiar language of spin). We cannot go too big with n, otherwise the running $SU(2)_L$ coupling explodes before the GUT scale. This is a somewhat arbitrary requirement, but if we buy into the assumption it turns out that $n \leq 5$ for fermions and $n \leq 6$ ($n \leq 4$) for real (complex) scalars [268].

Of course, the *dark* matter must be *dark*, so no electric charge Q. But this, for a given multiplet, automatically fixes the hypercharge assignment, since $Q = T_3 + Y$ (here of course T_3 indicates weak isospin).

Exercise 194. List all possible hypercharge values for $n = 2, 3, 4, 5$.

So: for every n we have 1 to 3 allowed values for the hypercharge, and two values of the spin. The mass M is then fixed for each case by a relatively straightforward calculation of the particle's thermal relic density (of course several of the multiplets have quasi-degenerate charged or doubly-charged partners, in which case co-annihilation, and a careful calculation of the spectrum, are necessary!). The result falls between about 0.4 TeV for the scalar doublet, up to 5 TeV for the $n = 5$ case [269], not accounting for non-perturbative effects (Sommerfeld enhancement, see Exercise 195).

At this point, to ensure stability, one should list all allowed operators (renormalizable or not) that could mediate decay of the dark matter. Take for example the $n = 2$ fermion. Hypercharge must be $1/2$, so the operator $\chi e H$, with e the (hypercharge -1) $SU(2)_L$-singlet ("right-handed") lepton field, and H the Higgs $SU(2)_L$ doublet (hypercharge $+1/2$) works fine, and would mediate $\chi \to eh$. And the decay would be very fast, so no stable dark matter! Doing the arguably tedious, but also somewhat exciting exercise of going through all possible such operators, one finds that in fact the list of "stable" minimal dark matter candidates is a relatively short one: we have got only $n = 5$ fermions, of hypercharge $Y = 0, 1$ or 2.

We saw in Chapter 4 that if $Y \neq 0$ there are *very* large contributions to the direct detection cross section, $\sigma \simeq G_F^2 m_p^2 Y^2$ (the first ones to estimate this were Goodman and Witten in 1985, [89]), which are vastly ruled out by current direct detection limits. Excitingly enough, we thus have only one single winner: $n = 5$, $Y = 0$, which happens to fix $M = 9.4$ TeV once the effect of $SU(2)_L$ bound states (Sommerfeld enhancement) are included.

Exercise 195. This exercise intends to guide you through what the calculation of the mentioned "Sommerfeld enhancement" looks like. Suppose there is an Abelian vector particle with mass M_V and gauge coupling α. At low energies, we will only be concerned with s-wave scattering. The non-perturbative correction we are after is given by

$$R = |\psi(\infty)/\psi(0)|^2,$$

where $\psi(r)$ is the reduced s-wave function for the two-body dark matter state, with energy K. The equation that ψ satisfies, in the non-relativistic limit, is just a good old Schrödinger equation,

$$-\frac{1}{M}\frac{d^2\psi}{dr^2} + V \cdot \psi = K\psi, \quad \text{with } V = \pm\frac{\alpha}{r}e^{-(M_V \cdot r)},$$

with the outgoing boundary condition $\psi'(\infty)/\psi(\infty) = iM\beta$ (it is a plane wave at infinity!), where $K = M\beta^2$. In the non-Abelian case of course this becomes a matrix equation, with V the sum of the various contributions, potentially including Higgs exchange.

Show that for $M_V \to 0$ one has the same Schrödinger equation as the hydrogen atom, which we learned to solve, analytically, in Kindergarten. Show that in that case

$$R = \frac{-\pi x}{1 - e^{\pi x}}, \qquad x = \pm\frac{\alpha}{\beta}.$$

This shows that R is sizable for $\beta \lesssim \pi\alpha$. Argue that this means that, at dark matter freeze-out, Sommerfeld enhancement (in the $M_V \to 0$ limit) is an $\mathcal{O}(1)$ effect.

This specific incarnation of "minimal" dark matter thus seems like a good target to shoot for. How could it be detected? Given how massive the new states are, one would need a much larger center-of-mass energy than those attainable at the LHC to probe this model (see e.g. [270,271]). However, one can hope to use direct and indirect dark matter detection. The direct detection story is rather interesting. In the model there is no spin-independent tree-level scattering of the dark matter candidate off of quarks, but it turns out that two-loop processes give as big an amplitude as one-loop processes.

Exercise 196. Explain why there are no tree-level contribution to the spin-independent cross section off of quarks; Sketch a couple of one-loop diagrams off of quarks and two-loop diagrams off of gluons, and argue that it is not crazy that those two classes of diagrams produce comparable contributions (see e.g. [272] for more details).

The bottom line is that future direct detection experiments will miss the predicted scattering cross section ($\sigma_{\rm SI}^{n} \simeq 2 \times 10^{-46}\ {\rm cm}^2$) just barely. Better news from indirect detection: the cross section for annihilation into $SU(2)_L \times U(1)_Y$ gauge bosons is calculable, and affected by significant "Sommerfeld" corrections, yielding at certain values for the mass very large pair-annihilation cross sections and abundant detectable final state particles (γ's, $\bar{p}$'s, e^+'s etc.). While the specific value $M = 9.4$ TeV seems to be accidentally unlucky, being close to a trough rather than a peak, I believe the jury is still out (do we really know how to accurately calculate the position of peaks and troughs in Sommerfeld enhancement? See e.g. Ref. [273] for the (similar) case of supersymmetric winos).

The inert doublet model

Another "minimal" extension that for a variety of reasons I like is the "inert (two-Higgs-) doublet" model (IDM)[274,275]: the key ingredient here is a second complex scalar $SU(2)_L$ doublet (in addition to the one in the SM), and an exact Z_2 symmetry under which (i) all SM particles and one of the scalar doublets are even, while (ii) the second scalar is odd. Additionally, all interaction terms in the model are posited to be renormalizable. The term "inert" derives from the fact that there is no coupling to SM particles involving one single particle in the second doublet.

Exercise 197. Denote the SM Higgs doublet with H and the "inert" doublet with Φ and write the most general tree-level scalar potential.

As a consequence of the Z_2 symmetry the lightest Z_2-odd particle (LOP) is stable and is a potential (WIMP-like) candidate for dark matter, as realized long ago [275].

Exercise 198. Calculate the scalar mass spectrum for the 5 scalars in the IDM given the potential you just wrote down in the exercise above.

Exercise 199. Argue that there cannot be any CP violation in the IDM.

The LOP is a prototypical WIMP dark matter candidate. At very low masses, the dominant annihilation mode is an s-channel Higgs exchange into whichever fermion pair is kinematically available; at half the SM Higgs mass this channel is resonant, and the coupling between the LOP and the Higgs (sometimes indicated with λ) can be quite small. Below the WW threshold, s-channel Z-exchange processes to fermion–anti-fermion pairs, albeit p-wave suppressed, are dominant [275]. Finally, at large LOP values, larger than about 100 GeV the annihilation to gauge boson pairs gets increasingly large, suppressing the relic density. However, if the heavy inert Higgses are all near-degenerate, a cancellation arises between the t/u inert Higgs exchange and the four-vertex amplitudes, and the relic density can again be close to the observed dark matter density.

Exercise 200. Sketch the relevant Feynman diagrams for the annihilation of IDM LOPs.

The most significant constraint on IDM dark matter comes from direct detection through Higgs exchange.

Exercise 201. Calculate the spin-independent cross section for the IDM LOP indicating with λ the coupling in the LOP–LOP–Higgs vertex.

Direct detection limits the viable LOP regions to (1) very low masses, (2) LOP masses close to half the Higgs mass (around 62–63 GeV or thereabout), or (3) large masses (larger than 0.5 TeV up to multiple TeVs).

Dark photons

One of the possible renormalizable "portals" between SM fields and new physics is kinetic mixing of a new, broken U(1)′ gauge sector with the SM hypercharge U(1), an operator that without loss of generality can be cast as

$$-\frac{\varepsilon}{2\cos\theta_W} B_{\mu\nu} F'^{\mu\nu}.$$

The gauge boson A' of the new Abelian force (whose field strength tensor is F') in the equation above has been called with a variety of names — from U-boson to hidden-sector, heavy-, para-photon; here, we will indicate it with the somewhat oxymoronic name of *dark photon*.

The kinetic mixing operator above induces an effective interaction between the electromagnetic current J^{μ}_{EM} and the dark photon A' (whose simple form explains the normalization factors employed above):

$$\varepsilon e A'_{\mu} J^{\mu}_{EM}.$$

The mixing parameter ε is in principle a free parameter of the theory, which is calculable, and naturally small, in specific model examples. The other free parameter is the mass $m_{A'}$ of the dark photon. Again, once a sector that breaks the U(1)′ is specified, such parameter is dynamically fixed. In many concrete incarnations, the mass is in the MeV–GeV range, but it can also naturally be much smaller, down to the meV (see Ref. [276] for a review). A simple generalization of this scheme is one where the dark photon mixes with the Z boson of the SM via mass mixing, and thus couples to both the electromagnetic and the weak neutral current of the SM, inducing interesting new possible phenomena such as "dark parity violation" etc. [276].

Searches for dark photons depend on the particle's mass, and they are substantially different for masses above and below $m_{A'} \sim 2m_e \sim 1$ MeV. "Heavy" dark photons can decay to pairs of electrically charged SM particles, and can be efficiently produced in electron or proton fixed-target experiments or at leptonic colliders [276]. If $m_{A'} < 1$ MeV, then (by Furry's theorem) only the decay to three photons is allowed.

Exercise 202.

(i) Estimate how the decay width of a dark photon to three photons scales with the relevant parameters α, ε, $m_{A'}$ and m_e.
(ii) Attach the "right" coefficient to the decay width [277],

$$\Gamma_{A' \to 3\gamma} = \frac{17}{11,664,000\pi^3} \alpha^{n_\alpha} \varepsilon^{n_\varepsilon} m_{A'}^{n_{m_{A'}}} m_e^{n_{m_e}},$$

and calculate the lifetime as a function of $m_{A'}$.
(iii) Argue that the 3-photon decay via the π^0 anomaly is suppressed by a high power (which power?) of $\Lambda_{\rm QCD}$, versus the (same) power of m_e [277].

Interestingly, as you just calculated, the lifetime of the dark photon is typically longer than the age of the universe as long as $m_A \lesssim 1$ MeV. As a result, one could contemplate dark photons as dark matter candidates — more below.

If the dark photon is very light it will oscillate into ordinary photons, and it can be searched for with techniques not dissimilar from those used to probe axion-like particles (ALPs) (see Chapter 7). Ultra-light dark photons (in the meV mass range) effectively behave like dark radiation, and could possibly be produced via oscillations with ordinary photons in just the right amount to explain extra light degrees of freedom possibly observed in the CMB if $\varepsilon \sim 10^{-6}$ [278]; this possibility is however strongly constrained by stellar evolution bounds [279], leaving dark radiation possibly being produced by other dark-sector particles' decays.

In general, the production of light, meta-stable dark photons with $m_{A'} \ll 2m_e$, which are potential dark matter candidates, proceeds through one of three mechanisms:

(i) photon-dark photon conversion, as mentioned above,
(ii) production through scattering or annihilation, analog to freeze-in, such as from processes like

$$\gamma + e^{\pm} \to A' + e^{\pm}, \qquad e^+ + e^- \to A' + \gamma,$$

(iii) production from an initial dark photon condensate (a rehash of the axion misalignment mechanism).

Exercise 203. Argue that the rate for mechanism (ii) above by dimensional analysis goes like

$$\Gamma_{(ii)} \sim \varepsilon^2 \alpha^2 \frac{n_e}{s},$$

where n_e is the density of electrons/positrons, and $\sqrt{s}$ is the center-of-mass energy of the process. Argue that at $T \ll m_e$ $\Gamma_{(ii)} \sim T$, and thus that most of the produced A' are produced at $T \sim m_e$.

It turns out that the only mechanism that could possibly produce dark photon dark matter is (iii), the parameter space compatible with the observed cosmological dark matter density for processes (i) and (ii) being ruled out by direct detection experiments [280], where cosmological dark photons ionize the target atoms. The relic dark photon density is estimated as [280]

$$\Omega_{A'} \sim 0.3\sqrt{\frac{m_{A'}}{1\ \mathrm{keV}}}\left(\frac{H_{\mathrm{inf}}}{10^{12}\ \mathrm{GeV}}\right),$$

where H_{inf} indicates the Hubble rate during inflation.

An interesting possibility that is slightly less "minimal" is that the dark photons are the vector fields of a *non-abelian* gauge theory. In that case, a kinetic mixing term is prohibited by gauge invariance automatically. A simple realization is a theory which contains this non-abelian dark photon and a complex scalar ϕ which acts as messenger with the SM through the Higgs $|H|^2$ portal interaction term, and which is charged under the non-abelian new symmetry but singlet under all SM gauge interactions. For example, if the new non-abelian symmetry is SU(2) one could have [281]

$$\mathcal{L} = \mathcal{L}_{\mathrm{SM}} - \frac{1}{4}F'^{\mu\nu} \cdot F'_{\mu\nu} + (D_\mu \phi)^\dagger((D^\mu \phi)) - \lambda|\phi|^2|H^2| - \mu_\phi^2|\phi|^2 - \lambda_\phi|\phi|^4,$$

with $D^\mu \phi = \partial^\mu \phi - i\frac{g_\phi}{2}\tau \cdot A'^\mu$. Since the scalar is in the fundamental representation of the gauge group, it features a custodial symmetry that implies that the 3 A'^μ gauge bosons are degenerate in mass and are automatically stable, without invoking any discrete symmetry! Ref. [281] explores the thermal relic density and direct detection prospects of this model.

Reference [79] studies a model where the non-Abelian gauge group is SU(3). This case exhibits several pieces of interesting phenomenology: for instance, it gives rise to prominent "semi-annihilation", i.e. the reaction $\psi_i\psi_j \to \psi_k\gamma$, and allows for numerous gamma-ray lines possibly enabling a sort of "dark sector" spectroscopy.

If the dark photon mediates interactions with a dark matter particle χ, then in principle it could also give rise to self-interactions that are in the interesting range $\sigma/m_\chi \sim 0.1-1\ \mathrm{cm}^2/\mathrm{g}$ (go back to Sec. 1.9 to refresh your memory of where this numbers come from, and to Sec. 9.4 for more on SIDM models). Approximating the dark photon-induced potential as a Yukawa potential, the resulting cross section is [282]

$$\sigma \sim 22.7/m_{A'}^2.$$

Exercise 204. Which hidden photon mass would make a 1 GeV dark matter particle sufficiently self-interacting to affect dark matter halos? How about for a 1 MeV particle?

Minimal concluding thoughts

The list of interesting, minimal dark matter models could go on for much longer. As Roderick Redman famously stated, "*Any competent theoretician can fit any given theory to any given set of facts*" [135]. If the theoretician is especially competent, the theory will also be minimal, in addition to explaining the "given set of facts". I am afraid in this day and age there is no shortage of competent theoreticians, nor of "anomalous" experimental or observational facts. Thus, on any given day the arXiv typically contains a plethora of "minimal" models, many with a dark matter candidate. Which makes any attempt at an exhaustive review, even in the form of a "laundry list", futile and quickly out-dated.

Appendix A

Rudimentary Particle Physics

A.1 The magic world of natural units

The system of natural units is defined by setting the speed of light in vacuum

$$c = 2.997\ldots \times 10^{10}\text{cm/s} = 1,$$

(thus giving distance and time the same units) and by setting Planck's "reduced" constant

$$\hbar = h/(2\pi) = 1.054\ldots \times 10^{-27}\text{erg s} = 6.582\ldots \times 10^{-22}\text{ MeV s} = 1,$$

effectively turning energy into inverse units of time. In natural units, length and time are *homogeneous* quantities,[a] with units of inverse energy or momentum; the action, velocity, and angular momentum are dimensionless quantities, etc. The system is especially convenient when dealing with a relativistic (c), quantum ($\hbar$) theory, for example *quantum field theory*. In addition, one can set the Boltzmann constant

$$k_B = 1.381\ldots \times 10^{-16}\text{ erg/K} = 1,$$

effectively translating units of energy into temperature, and vice versa — something useful here, e.g. in applications of statistical mechanics to cosmology.

In natural units all physical quantities are measured in units of one single entity, for example energy (or mass), to some power. Indeed, energy

[a]Same units, that is.

(e.g. GeV) is, usually, the unit of choice in particle physics. Velocity has units of energy to the power zero, distances and times of energy to the power -1, (space) derivatives and momenta to the power $+1$, Lagrangian densities to the power $+4$, etc.

Exercise 205. Calculate the mass dimension of a real scalar field ϕ and of a spin 1/2 field ψ using the fact that the action is a dimensionless quantity, and knowing that the kinetic terms look like $(\partial_\mu \phi)^2$ and $\bar{\psi}\partial_\mu \gamma^\mu \psi$ (Incidentally, this generalizes to fermionic and bosonic fields of any spin).

It is important to be able to "reconstruct normal units" back from this magic world of natural units, and, sometimes, vice versa. This is done by using the only possible appropriate combination of powers of c and $\hbar$. For example, to turn a cross section[b] in natural units (energy to the power -2) back into "normal" units of area, we need to put back the right factors of c and $\hbar$, in this case $(\hbar c)^2$; to turn a lifetime, defined as an inverse decay width $1/\Gamma$ (units of energy to the power -1) back into units of time we need a factor $\hbar$, i.e. $\tau\ [s] = \hbar/\Gamma$. There are a couple conversion constants that appear often, and which in convenient units read

$$\hbar c = 197.3\ldots \mathrm{MeV fm},$$

$$(\hbar c)^2 = 0.389\ldots \mathrm{GeV}^2\ \mathrm{mbarn},$$

where 1 barn $\equiv 10^{-24}$ cm^2 (The name barn has an interesting story, intertwined with the Cold War: looking for a secretive name for the unit to describe the cross-sectional area of a typical nucleus, an "easy" target to strike for particle accelerators, American physicists at Purdue University adopted the name barn from the colloquial expression "could not hit the broad side of a barn"; this expression, incidentally, applies well to the Author, who most definitely cannot play baseball).

Natural units also allow to extract "natural" physical scales from fundamental constants. A prime example is the definition of Planck mass M_P, defined as Newton's gravitational constant G_N in natural units, to a power such that it has (natural) units of energy (i.e. in natural units, mass).

---

[b]See the next section for a definition of scattering cross section and decay width.

Exercise 206. Show that $M_P \sim G_N^{-1/2}$ and calculate it in GeV units, converting from $G_N = 6.674\ldots \times 10^{-11}$ m^{3}kg^{-1}s^{-1}.

Exercise 207. Calculate a typical gravitational-interaction cross section $\sigma \sim 1/M_P^2$ in barn. Calculate the Planck length $1/M_P$ in cm, and discuss its physical significance.

A.2 Cross sections and decay widths

Consider a (uniform, for simplicity) beam of particles of type 1, of mass m_1, with number density n_1 and velocity v_1 impinging on a target of particles of type 2, mass m_2, of (uniform) number density n_2 at rest. Given the forces acting between particles of type 1 and 2, an interaction, or scattering, has a certain probability to occur. Such probability, or equivalently and more quantitatively the number of scattering events dN that occurs per unit time and unit volume, is proportional to the incoming "flux" $v_1 n_1$ and to the density n_2. The proportionality constant σ is called (and defines the) "scattering cross section":

$$dN = \sigma(v_1 n_1) n_2 dV dt.$$

As you can see,[c] the cross section has dimensions of an area (inverse energy squared in natural units). Any quantum field theory textbook walks you through the appropriate definitions and normalizations of in- and out-states, and how to cast the expression above in a Lorentz-invariant way (i.e. how to write *the* Lorentz-invariant quantity that in the rest frame of particles 2 is given by the expression above, see e.g. Ref. [283]) — let me just summarize the bottom line here.

Generically, a *differential* cross section is written as the product of the final state Lorentz-invariant phase-space factor for a generic n-particles final state,

$$d\Phi^{(n)} = (2\pi)^4 \delta^{(4)}(P_i - P_f) \prod_{i=1}^{n} \frac{d^3 p_i}{(2\pi)^3 (2E_i)},$$

times the appropriate matrix element squared $|\mathcal{M}_{fi}|^2$, with $P_{i,f}$ the initial and final total four-momenta, divided by a flux factor.

[c]And if you cannot please do the little exercise!

Exercise 208. Show that $d\Phi^{(n)}$ is a Lorentz-invariant quantity, and that each factor d^3p_i/EA_i is separately Lorentz-invariant.

The latter can be cast in different (more or less convenient) forms. The one I like, which is manifestly Lorentz-invariant, is

$$I = \sqrt{(p_1 \cdot p_2)^2 - m_1^2 m_2^2} = E_1 E_2 \sqrt{(\vec{v}_1 - \vec{v}_2)^2 - (\vec{v}_1 \times \vec{v}_2)^2}. \qquad \text{(A.1)}$$

Exercise 209. Demonstrate the last equality in Eq. (A.1).

With that choice for the flux factor, the differential cross section reads

$$d\sigma = \frac{1}{4I}|\mathcal{M}_{fi}|^2 d\Phi^{(n)}, \qquad \text{(A.2)}$$

where $\mathcal{M}_{fi}$ is the transition amplitude (in the quantum mechanical sense, i.e. the matrix element) between the initial and final state. If the initial state particles are co-linear, the flux factor $I = E_1 E_2 |\vec{v}_1 - \vec{v}_2|$, which is more commonly seen in the definition of cross section, but not as generic as the expression in Eq. (A.1).

The expression in Eq. (A.2) allows you to calculate cross sections for generic processes, as long as you have an idea of how to calculate the matrix element squared $|\mathcal{M}_{fi}|^2$: it is usually simple to estimate the flux factor I and the phase-space factor (see e.g. the handy formula for a two-body final state below). Thinking about it, and using these three separate elements allows you to make a better job than employing simple dimensional analysis (i.e. improving on arguments like: "hey, it is a cross section, dimensions mass to the minus two, this is the relevant mass scale, bam!").

The decay rate $d\Gamma$ is defined as the probability dP that a single particle decays to a generic n-particles final state over a time T,

$$d\Gamma = \frac{1}{T} dP,$$

and has a similar structure to the cross section. Quantum mechanics dictates that such probability is the product of the same final-state Lorentz-invariant phase-space we introduced above, times, again, the appropriate matrix element squared. The correct relativistic normalization of single-particle states then gives

$$d\Gamma = \frac{1}{2E_p}|\mathcal{M}_{fi}|^2 d\Phi^{(n)},$$

with E_p the decaying particle's energy. The total decay width $\Gamma = \int d\Gamma$ is the inverse of the particle's lifetime.

Exercise 210. Prove that for $n = 2$ integration over the delta function enforcing conservation of total energy-momentum leads to the following differential phase-space, for the decay of a particle of mass M into two particles of mass $m_{1,2}$:

$$d\Phi^{(2)} = \frac{1}{32\pi^2}\left[M^4 + (m_1^2 - m_2^2)^2 - 2M^2(m_1^2 + m_2^2)\right]^{1/2} d\Omega.$$

Rewrite the formula above for couple of interesting limiting cases:

(i) $m_1 = m_2 = m$, and
(ii) $m_1 = 0$ and $m_2 = m$.

A.3 Effective theories in a nutshell

Matthew Schwartz, in Chapter 22 of his excellent book, Ref. [284] (on which this section is largely based), quips that "not all of quantum field theory consists of computing loops in renormalizable theories" (incidentally: thank goodness!). Effective theories are non-renormalizable theories with a limited, more or less well-defined range of applicability, that are often *more useful* than their full, renormalizable counterparts within that range.[d]

Technically, a non-renormalizable interaction is one where the *mass dimension* of the coefficient multiplying the quantum fields participating in the interaction *is negative*. Interestingly, non-renormalizable theories *can be renormalized*, at the price, however, of having to add (to all orders) an *infinite number* of counter-terms. This does not make them, however, less predictive or less useful.

In his chapter, Schwartz mentions four examples of effective theories, one per "fundamental interaction". The first example is Schrödinger's

[d]Note that non-renormalizable does not mean that one cannot calculate loops yielding finite, testable physical results! A classic example is the non-analytic scale dependence $\ln \frac{s_2}{s_1}$.

equation,

$$i\partial_t \psi = \left(-\frac{\nabla^2}{2m} + V(r)\right)\psi,$$

which you can regard as the "effective theory" (in the non-relativistic limit) of the Dirac equation, with the mass-dimension -1 parameter $1/m$. Consistency requires adding any term allowed by the symmetries of the Hamiltonian, that is

$$H = \frac{\vec{p}^2}{2m}\left(1 + a_1\frac{\vec{p}^2}{m^2} + a_2\frac{\vec{p}^4}{m^4} + \cdots\right) + V(r),$$

with a_i coefficients that could in principle be calculable from matching the "UV-complete" theory, here Dirac's equation, or, in the range of validity of the theory ($|\vec{p}| \ll m$) determined from experiments (and reused to make predictions, again in the effective theory's range of validity, for other physical observables).

Exercise 211. Calculate the first two coefficients a_1 and a_2 from Dirac's equation; estimate the effects of the a_1 term on the energy levels of the hydrogen atom in quantum mechanics. Compare with those from a $\ln\frac{\vec{p}^2}{m^2}$ term — which would be easier to measure?

A famous and very useful effective theory we use extensively in this book is Fermi's four-fermion interaction, originally introduced by Italian[e] physicist Enrico Fermi in 1933 to describe β-decay,

$$\mathcal{L}_{\text{Fermi}} = G_F(\bar{\psi}_p\Gamma\psi_n)(\bar{\psi}_e\Gamma\psi_\nu),$$

with the symbol Γ indicating some possibly non-trivial Dirac gamma-matrix structure. Here, the coupling of the effective theory, Fermi's constant G_F, has mass dimension -2 (physically, the corresponding mass scale is associated with the mass of the W boson of weak interactions, which is integrated out at low energy in the effective theory). Higher-order terms here are of the form $(G_F E^2)^j$, $j \geq 1$. The theory makes sense without those terms as long as $(G_F E^2) \ll 1$, i.e. if the energy scale of the process

[e]Or, as engraved at the entrance of Fermilab, the "American Physicist, born Italian". Whatever.

$E \ll 1/\sqrt{G_F} \sim m_W$. This energy scale sets the *range of applicability* of the effective theory.

Note that the Fermi Lagrangian above is quite predictive: for example, it can be used to relate processes like $n \to p^+e^-\bar{\nu}$ to $p^+ \to ne^+\nu$; it provides predictions for the angular and energy distribution of the beta-decay products etc. Also, one can probe parity violation by considering different γ matrix structures in the spinor bilinears, for example $\bar{\psi}\gamma_5\gamma^\mu\psi\bar{\psi}\gamma_5\gamma_\mu\psi$.

The same effective theory was advocated to describe other weak interaction processes, such as muon decay,

$$\mathcal{L}_{\mathrm{Fermi},\mu} = \frac{G_F}{\sqrt{2}}\left(\bar{\psi}_\mu\gamma_\lambda(1-\gamma_5)\psi_{\nu_\mu}\right)\left(\bar{\psi}_e\gamma^\lambda(1-\gamma_5)\psi_{\nu_e}\right).$$

Muon decay, incidentally, provides one of the best measurements of the Fermi constant,

$$G_F = 1.13337(1) \times 10^{-5}\ \mathrm{GeV}^{-2}.$$

In principle, at the effective theory level, there is no reason why the G_F of β-decay and of muon decay are the same — but the fact that they are is a powerful hint for the "full-glory" UV-complete theory. The effective theory should include all terms allowed by symmetry — disregarding the various fermion bilinear γ matrix structures, and suppressing fermion species subscripts, one should include terms such as

$$\mathcal{L} = G_F\bar{\psi}\psi\bar{\psi}\psi + a_1G_F^2\bar{\psi}\psi\Box\bar{\psi}\psi + a_2G_F^2\bar{\psi}\partial\psi\Box\bar{\psi}\partial\psi + \cdots.$$

The terms on the right are unimportant as long as the typical energy E of the process is small enough compared to the Fermi coupling energy scale, i.e. if $G_FE^2 \ll 1$.

As the Reader knows, what we think is *the* UV completion of the Fermi theory is the theory of electroweak interactions, based on the non-abelian gauge group SU(2)×U(1). To simplify matters, consider, after spontaneous symmetry breaking, a theory with one massive vector boson W_μ interacting with fermions,

$$\mathcal{L} = -\frac{1}{4}F_{\mu\nu}^2 + \frac{1}{2}M^2W_\mu^2 + \bar{\psi}(i\partial + gW)\psi, \quad F_{\mu\nu} \equiv \partial_\mu W_\nu - \partial_\nu W_\mu.$$

This is of course a simplification — the actual piece of the electroweak Lagrangian involving the W boson is more complicated, but bear with me. The matrix element for a $\psi\psi \to \psi\psi$ scattering is, from basic Feynman rules,

$$i\mathcal{M} \sim (ig)^2 \bar{v}_2 \gamma^\mu u_1 \frac{-i\left(g^{\mu\nu} - \frac{p^\mu p^\nu}{M^2}\right)}{s - M^2} \bar{u}_3 \gamma^\nu v_4,$$

where the $\frac{p^\mu p^\nu}{M^2}$ vanishes (in this simple U(1) approximation) by the Ward identity. At $s \ll M^2$ the matrix element reduces to

$$i\mathcal{M} = -i\frac{g^2}{M^2} \bar{v}_2 \gamma^\mu u_1 \bar{u}_3 \gamma_\mu v_4.$$

Physically, the massive gauge boson propagates over distances $\sim M^{-1}$ much shorter than those probed by the energies of the scattering process, $\sim(\sqrt{s})^{-1}$. The "effective" matrix element also allows the matching onto the effective theory, giving

$$G_F = \frac{g^2}{M^2}.$$

In reality, the expression in the SU(2)×U(1) theory is a bit more complicated, $G_F = \frac{\sqrt{2}}{8}\frac{g_w^2}{m_W^2}$, with $m_W \simeq 80.4$ GeV and $g_w \simeq 0.65$. A power series expansion in s/M^2 gives successive higher-order terms for the effective theory.

Exercise 212. Calculate the term of order M^{-4}, using on-shell spinors.

A final example of effective theory is the Chiral Lagrangian, a very powerful construction describing low-energy interactions of light mesons and hadrons. This can be relevant for theories of light dark matter particles ($m_X < 1$ GeV) interacting, at the fundamental level, with quarks but, effectively, with the physical degrees of freedom at those energies, i.e. light mesons and hadrons [285].

The Chiral Lagrangian hinges on the notion that the strong nuclear force is invariant under (strong) isospin, an SU(2) symmetry under which the proton and the neutron transform as a doublet $\psi_i = (p, n)$. Of course, electromagnetic (and weak) interactions violate isospin. Pions are, in this framework, particles that "mediate" interactions between hadrons such as the proton–neutron doublet, and must thus respect the isospin SU(2) symmetry. Pions are accommodated in a single matrix, using Pauli's σ matrices, in the following way:

$$U(x) = \exp\left[\frac{i}{F_\pi}\begin{pmatrix} \pi^0(x) & \sqrt{2}\pi^-(x) \\ \sqrt{2}\pi^0(x) & -\pi^0 + (x) \end{pmatrix}\right] = \exp\left[\frac{i}{F_\pi}\sigma^a\pi^a(x)\right],$$

with the identifications

$$\pi^0 = \pi^3, \quad \pi^\pm = \frac{1}{\sqrt{2}}\left(\pi^1 \mp i\pi^2\right),$$

and with $F_\pi \simeq 92$ MeV the pion decay constant. The name *chiral* and the odd form in which the pions are represented come from the so-called (spontaneously broken) chiral $SU(2)_L \times SU(2)_R$ symmetry acting on left- and right-handed quarks in the quantum chromodynamics (QCD) Lagrangian, of which the pions are the corresponding Goldstone bosons. The Chiral Lagrangian is then the collection of possible terms that respect the (isospin) SU(2) symmetry, the simplest of which involving U is

$$\mathcal{L}_\chi = \frac{F_\pi^2}{4}\mathrm{Tr}\left[(D_\mu U)(D_\mu U)^\dagger\right] + \cdots,$$

where $D_\mu = \partial_\mu - iQ_iA_\mu$ is the standard quantum electrodynamics (QED) covariant derivative. The Lagrangian term above, expanded in terms of the pion fields $\pi^0, \pi^\pm$ gives, to lowest order in powers of F_π^{-1},

$$\mathcal{L}_{\mathrm{kin}} = \frac{1}{2}(\partial_\mu\pi^0)(\partial_\mu\pi^0) + (\partial_\mu\pi^+)(\partial_\mu\pi^-),$$

which are just the kinetic terms, and photon interactions for the charged pions. Higher-order terms give interaction terms such as

$$\mathcal{L}_{\mathrm{int}} = \frac{1}{F_\pi^2}\left(-\frac{1}{3}(\pi^0\pi^0)\partial_\mu\pi^+\partial_\mu\pi^- + \cdots\right) + \frac{1}{F_\pi^4}\left(-\frac{1}{18}(\pi^-\pi^+)^2\partial_\mu\pi^0\partial_\mu\pi^0 + \cdots\right) + \cdots,$$

which explicitly shows that, indeed, the Chiral Lagrangian is an effective (non-renormalizable) theory.

Exercise 213. Calculate the numerical coefficients terms for the Chiral Lagrangian of order F_π^{-2} and F_π^{-4} in the equation above, and give examples of a few additional terms.

The breakdown of the validity of the Chiral Lagrangian occurs when $\sqrt{s} \sim 4\pi f_\pi \sim 1200$ MeV where $f_\pi \simeq 93$ MeV is the pion decay constant. Above that scale the theory is no longer predictive, i.e. useful, and the correct theory of strong nuclear interactions is QCD, the theory of quarks and gluons. Perturbative QCD has no pions at all — these are non-perturbative low-energy bound states which arise e.g. in non-perturbative treatments of QCD such as lattice. At low energies, the Chiral Lagrangian is infinitely more useful than QCD in describing strong nuclear interactions. However, as for any effective theory, such usefulness is confined to the effective theory's range of validity.

A.4 A succinct primer on neutralino dark matter

There are numerous excellent reviews[f] on supersymmetry and on low-energy supersymmetric extensions of the Standard Model (SM) (such as the minimal supersymmetric extension of standard model (MSSM)), which make an even succinct synthesis rather superfluous. What I intend to do here is, instead, to give a short compendium of the properties of neutralinos as dark matter particles, and how those properties depend on parameters in the MSSM.

In the absence of right-handed neutrino superfields, neutralinos are the only viable dark matter candidates in the MSSM that can be produced thermally: thermal gravitinos do not work, as discussed in Sec. 9.2, and (left-handed) sneutrinos can only be the lightest supersymmetric particles (LSP) if they are very heavy, due to direct detection constraints (see Sec. 4 for an estimate of the Z boson exchange direct detection cross section), in a mass range where thermal production is excluded.

Neutralinos are the mass eigenstates of the fermionic supersymmetric partners of neutral electroweak gauge bosons (the "gauginos": $\tilde{B}$, also known as bino, the fermionic partner of the hypercharge gauge boson, and $\tilde{W}_3$, the "wino", the partner of the W^0 SU(2) neutral gauge boson) and of the (two) neutral Higgses of the MSSM (the "higgsinos" $\tilde{H}_i^0$, $i = 1, 2$). Ironically, the suffix "-ino" in Italian means "the smaller version of", while here of course the -inos are usually much heavier than their counterparts.

[f]See e.g. Refs. [223,286,287] and the review articles [100,288].

After supersymmetry and electroweak symmetry breaking, the four interaction eigenstates mix, the relevant mass matrix, in the $(\tilde{B}, \tilde{W}_3, \tilde{H}_1^0, \tilde{H}_2^0)$ basis being

$$\mathcal{M}_{\tilde{\chi}^0} = \begin{pmatrix} M_1 & 0 & -\frac{g' v_1}{\sqrt{2}} & +\frac{g' v_2}{\sqrt{2}} \\ 0 & M_2 & +\frac{g v_1}{\sqrt{2}} & -\frac{g v_2}{\sqrt{2}} \\ -\frac{g' v_1}{\sqrt{2}} & +\frac{g v_1}{\sqrt{2}} & 0 & -\mu \\ +\frac{g' v_2}{\sqrt{2}} & -\frac{g v_2}{\sqrt{2}} & -\mu & 0 \end{pmatrix}, \tag{A.3}$$

with $v_{1,2}$ the vacuum expectation value of the real, neutral components of the two SU(2) Higgs doublets, and g and g' the SU(2) and U(1) hypercharge gauge coupling constants. The relevant mass parameters are the soft supersymmetry breaking U(1) and SU(2) gaugino masses M_1 and M_2, and the supersymmetric Higgs–higgsino mass parameter μ. Given the matrix above, one can find, in the standard way, eigenvalues, $m_{\tilde{\chi}_i^0}, i = 1, \ldots, 4$, and eigenvectors,

$$\tilde{\chi}_i^0 = N_{i1}\tilde{B} + N_{i2}\tilde{W}_3 + N_{i3}\tilde{H}_1^0 + N_{i4}\tilde{H}_2^0,$$

where we have chosen the convention where the matrix N_{ij} is complex-valued and the mass eigenvalues are all positive (another popular convention has the matrix N be real and orthogonal, which however implies that some of the eigenvalues can be negative; this is taken care of by a chiral rotation in the field definition for neutralinos with negative eigenvalues). Note that the mass matrix above should also contain radiative corrections (again, see dedicated reviews on this, e.g. [289]). Let me just point out that the most important ones affect the (zero, at tree-level) entries (3,3) and (4,4) of the mass matrix above and stem from one-loop corrections from (s)bottom and (s)top quarks.

The composition of the lightest neutralino is key to its phenomenology as a dark matter particle, and is driven by the relative magnitude of the parameters M_1, M_2, and μ. The couplings and properties of a mass eigenstate that is very close to one of the interaction eigenstates $\tilde{B}, \tilde{W}_3, \tilde{H}_1^0, \tilde{H}_2^0$ are simple and easy to predict based on SM interaction vertices. It is arbitrary but somewhat useful to define a bino, wino and higgsino fraction for the lightest neutralino as:

$$\begin{aligned}\text{bino frac.}: &\ |N_{11}|^2,\\ \text{wino frac.}: &\ |N_{12}|^2,\\ \text{higgsino frac.}: &\ |N_{13}|^2+|N_{14}|^2.\end{aligned}$$

The sum of all of the -ino fractions is one, and sometimes people use the terms "higgsino", "wino" or "bino"-like lightest neutralino if the corresponding fraction is larger than some value, say 0.90 or 0.99, and talk about "mixed neutralinos" in the other cases.

To quickly appreciate how the different -ino fractions dictate important aspects of the phenomenology, let me make a couple of examples. In many instances, the lightest neutralino spin-independent cross section is driven by t-channel CP-even Higgs exchange. The key coupling driving the magnitude of the scalar cross section off of quarks is then the neutralino–neutralino–CP-even Higgs(es) coupling. That coupling is, in turn, approximately proportional to the product of the gaugino (bino plus wino) times the higgsino fraction — it is thus suppressed for "pure" neutralino states, and enhanced for maximal mixing. The coupling to the Z boson, instead, drives spin-dependent interactions, and is proportional to the slightly less trivial combination $[|N_{13}|^2-|N_{14}|^2]$ (times an SU(2) factor $g/(2\cos\theta_W)$), which has to do with (but is not!) the higgsino content (in fact the coupling identically vanishes for pure higgsino states!).

A second sector that is often, if not almost always, relevant to the phenomenology of neutralinos as dark matter candidates is the *chargino* sector. The relevant mass matrix is here

$$\begin{pmatrix}\tilde{W}^- & \tilde{H}_1^-\end{pmatrix}\mathcal{M}_{\tilde{\chi}^\pm}\begin{pmatrix}\tilde{W}^+\\ \tilde{H}_2^+\end{pmatrix}+\text{h.c.},$$

where the mass matrix

$$\mathcal{M}_{\tilde{\chi}^\pm}=\begin{pmatrix}M_2 & gv_2\\ gv_1 & \mu\end{pmatrix},$$

is diagonalized by

$$\begin{aligned}\tilde{\chi}_i^- &= U_{i1}\tilde{W}_i^-+U_{i2}\tilde{H}_1^-,\\ \tilde{\chi}_i^+ &= V_{i1}\tilde{W}_i^++U_{i2}\tilde{H}_2^+.\end{aligned}$$

One can choose det(U)=1 and $U^* \mathcal{M}_{\tilde{\chi}^\pm} V^\dagger = \mathrm{diag}(m_{\tilde{\chi}_1^\pm}, m_{\tilde{\chi}_2^\pm})$ with non-negative mass eigenvalues. It is simple enough to derive analytic forms for the U and V matrices (see e.g. Ref. [286]).

Exercise 214. Show that

$$\sum_{i=1,2} m_{\tilde{\chi}_i^\pm} = \mu^2 + M_2^2 + 2M_W^2$$

and that

$$\sum_{i=1,\ldots,4} m_{\tilde{\chi}_i^0} = 2\mu^2 + M_1^2 + M_2^2 + 2M_Z^2$$

(remember that $M_W^2 = \frac{g^2}{2}(v_1^2 + v_2^2)$ and $m_Z^2 = \frac{g^2+g'^2}{2}(v_1^2 + v_2^2)$). (This and the following three exercises are from the excellent book by Baer and Tata, Ref. [286], Sec. 8.3.)

Exercise 215. Suppose the soft supersymmetry breaking masses $M_{1,2} = 0$. Show that the lightest neutralino is then a massless "photino" (think what the name means) and that the "wino" and "zino" are both lighter than M_W and M_Z, respectively. Show that if $M_1 = M_2 = M$ the same photino you constructed above is an eigenstate of the neutralino mass matrix with mass M, at tree level.

Exercise 216. Show that the lightest neutrino is a massless higgsino if $\mu = 0$, and calculate it as a linear combination of $\tilde{B}, \tilde{W}_3, \tilde{H}_1^0, \tilde{H}_2^0$ using $\tan\beta = v_2/v_2$.

Exercise 217. Show that the lightest neutralino is massless if

$$\mu^2 + M_W^2 \sin 2\beta \left(\frac{1}{M_2} + \frac{\tan^2\theta_W}{M_1} \right) = 0$$

and find the corresponding eigenvector.

As you can see by eye, in general higgsino-like -inos come with a set of quasi-degenerate two neutralinos and one chargino,[g] all close in mass to μ, while winos come with one neutral and two charged guys with mass $\sim M_2$. The near-mass degeneracy in the higgsino and wino systems *always* implies important *co-annihilation* effects in those systems (see the discussion in Sec. 3.3).

In the chargino–neutralino system the coupling to the Z and W gauge boson is key to whether or not pair-annihilation to massive gauge boson pairs is important in setting the thermal relic abundance.[h]

The coupling to the Z is as above (proportional to $[|N_{13}|^2 - |N_{14}|^2]$), while the other important coupling is the chargino–neutralino–W. Let me give the explicit form for the latter, in terms of the chiral projectors $P_{R,L} = (1 \pm \gamma_5)/2$:

$$g_{W^\pm \tilde{\chi}_i^0 \tilde{\chi}_j^\mp} = \gamma^\mu \left[\left(N_{i2} V_{j1}^* - \frac{N_{i4} V_{j2}^*}{\sqrt{2}} \right) P_L + \left(N_{i2}^* U_{j1} - \frac{N_{i3}^* U_{j2}}{\sqrt{2}} \right) P_R \right].$$

The coupling is thus a non-trivial combination of the chargino and neutralino mixing matrices. It is always large for wino-like and for higgsino-like neutralinos, or mixed states thereof.

The spectrum of other supersymmetric particles is also of key importance for the phenomenology of neutralinos as dark matter particles. One example is resonant s-channel Higgs exchange, which, if $m_{\tilde{\chi}_1^0} \simeq m_S/2$, with S one of the neutral Higgs states, can dominate the neutralino pair-annihilation cross section both in the early universe and, if $S = A$, see Exercise 218, at late times, driving the annihilation final states to whichever final states S decays into.

Exercise 218. Explain why the pair-annihilation cross section of two Majorana fermions via the exchange of a CP-even scalar vanishes at zero relative velocity (s-wave).

Another example is a spectrum where one sfermion (say, with a wee-bit of theoretical prejudice, a stau or a stop) is almost degenerate with the lightest neutralino. In this case co-annihilation is important and if, at $T = 0$, the sfermion t-channel exchange dominates the pair-annihilation

[g]Of course there is really two exactly degenerate charginos, $\pm$.

[h]Of course, as long as the lightest neutralino is heavier than the weak gauge bosons, i.e. if the weak gauge bosons are kinematically allowed in the final state.

cross section, the corresponding fermion-pair final state will dominate the annihilation modes.

Exercise 219. Explain why the pair-annihilation cross section of two Majorana fermions at zero relative velocity into a pair of SM fermions of mass m_f is proportional to m_f^2. Suggestion: Think about what happens with charged pion decay.

A.5 A succinct primer on (minimal) universal extra dimension's (UED) Kaluza–Klein dark matter

For the Reader eager to know everything about Kaluza–Klein dark matter and more, see the brilliant review by Dan Hooper and Yours Truly, Ref. [64]. For the not-so-eager Reader, here is a much, much shorter account.

Suppose there existed an extra spatial dimension[i] and all SM fields were allowed to propagate in the extra dimension, which is compactified on a circle[j] of radius R. Cleverly choosing how fields transform under $y \to -y$, where y is the fifth dimensional coordinate, appropriately kills the zero-modes of unwanted fields, for example the gauge field component A_μ for $\mu = 5$.

Exercise 220. Show that you can decompose scalar fields which are even (H) and odd (A_5) under $y \to -y$ as

$$H(x^\mu, y) = \frac{1}{\sqrt{\pi R}}\left[H_0(x_\mu) + \sqrt{2}\sum_{n=1}^{\infty} H_n(x_\mu)\cos\left(\frac{ny}{R}\right)\right];$$

$$A_5(x^\mu, y) = \sqrt{\frac{2}{\pi R}}\sum_{n=1}^{\infty} A_{5,n}(x_\mu)\sin\left(\frac{ny}{R}\right).$$

Infer from this expression the (tree-level) mass spectrum for the n-th mode in the effective four-dimensional theory.

[i] All of the discussion generalizes, more or less painfully, to more than one extra dimension, but for simplicity I restrict this primer to only one.

[j] In fact, requiring chiral fermions in the four-dimensional theory forces one to have special boundary conditions, namely to compactify the extra dimension on the orbifold S^1/Z_2.

As you have explicitly calculated above, the tree level mass of the n-th Kaluza–Klein excitation of a SM field $X^{(n)}$ is

$$m^2_{X^{(n)}} = \frac{n^2}{R^2} + m^2_{X^{(0)}},$$

where the latter term is the SM mass. For large n, or for small $m^2_{X^{(0)}} \ll 1/R$, $m_{X^{(n)}} \simeq n/R$, and the spectrum appears like a tower of equally-spaced and quasi-degenerate particles for every n. The tree-level mass gets radiative corrections that are crucial for phenomenology, and set the nature and properties of the "lightest Kaluza–Klein particle" (LKP). Such radiative corrections depend on the theory cutoff scale, Λ, which is a parameter of the theory. The largest value of Λ is set by requiring the theory to have perturbative five-dimensional gauge couplings (which implies $\Lambda R \sim 10$) or vacuum stability (which sets a much stronger upper limit, but is somewhat model-dependent).

One might naively hope KK number to be conserved, since the KK number is a measure of the momentum of the particle in the extra dimension. Unfortunately, the (phenomenologically necessary) orbifold boundary conditions break translational symmetry in the extra-dimension. However, a subgroup of the KK number, known as KK parity, remains intact, and ensures conservation of the "evenness" or "oddness" of KK number at a vertex. KK parity is key to LKP's stature as a dark matter candidate, since it enforces its stability.

Before the discovery of the Higgs, the LKP could have been either the Kaluza–Klein hypercharge gauge boson $B^{(1)}$, the Kaluza–Klein graviton $G^{(1)}$ or the Kaluza–Klein Higgs $H^{(1)}$ (for very large SM Higgs masses, the LKP can even be charged, which, the LKP being stable, would have been automatically excluded). The measured SM Higgs mass, $m_h \simeq 125$ GeV, has settled this question, at least under the "minimal" assumption that so-called boundary kinetic terms vanish at the cutoff scale Λ: for $1/R \lesssim 800\,\text{GeV}$ the LKP is the Kaluza–Klein graviton $G^{(1)}$, for larger $1/R$ the LKP is the $B^{(1)}$. The KK graviton as an LKP is a "super-WIMP", produced in the early universe, out of equilibrium, by processes similar to those producing gravitinos in supersymmetry (see Sec. 9.2). The phenomenology of KK gravitons as LKPs depends crucially on the mass splitting $\Delta m = m_{B^{(1)}} - m_{G^{(1)}}$ which, for the minimal model, is always smaller than a couple GeV.

Exercise 221. Estimate the KK graviton decay width, and argue that

$$\Gamma(B^{(1)} \to G^{(1)} + \gamma) \sim \frac{(1/R)^3}{M_4^2}(\Delta m)^3,$$

where $M_4^2 \equiv (16\pi G_N)^{-1}$.
Estimate the KK graviton lifetime for $\Delta m \sim 1$ GeV.

For Δm set by the minimal model boundary conditions, the $G^{(1)}$ is long-lived enough that late decays are excluded both by constraints from the diffuse photon spectrum and from distortions to the cosmic microwave background (CMB) blackbody spectrum, if the KK graviton is the dark matter. In the minimal model, therefore, the LKP of interest is the $B^{(1)}$, a prototypical spin 1 weakly interacting massive particles (WIMPs).

The thermally averaged pair-annihilation cross section of $B^{(1)}$'s is roughly temperature independent, and for obvious dimensional reasons the pair-annihilation cross section scales as $\sigma \sim g_1^4/(R^{-1})^2$. In setting the ($B^{(1)}$) LKP relic abundance, co-annihilation processes (see Sec. 3.3) are critically important. Also, given the spectrum of the KK modes, resonant annihilation through level-2 KK modes s-channel exchange is also crucially important (again, see Sec. 3.3). The number of processes involved, and the quasi-automatic resonant structure of the cross sections makes this a pretty interesting and not-so-simple problem. The most up-to-date calculations for the LKP thermal relic abundance are given in Ref. [153], and indicate that, depending on the cutoff scale Λ, the inverse radius that produces a thermal relic density of $B^{(1)}$'s in accordance with the density of cold dark matter is between 1.2 and 1.4 TeV (see Fig. A.1). Our study also shows that (i) current collider and direct detection searches fall short of probing that compactification scale but that (ii) future searches will thoroughly close in on it (Hurray for testable models!).

Unlike in the case of Majorana fermions, the $f\bar{f}$ final state is not helicity-suppressed for spin 1 WIMPs, and, for the $B^{(1)}$ the branching fraction depends simply on the hypercharge of the fermion f (to the fourth power). This has important implications for indirect searches. For gamma-ray searches, the radiative process $B^{(1)}B^{(1)} \to f\bar{f}\gamma$ is also very important.

> **Exercise 222.** Estimate the ratio between the cross section for $B^{(1)}B^{(1)} \to f\bar{f}\gamma$ and $B^{(1)}B^{(1)} \to f\bar{f}$. See Ref. [290] for the full solution.

The spin-independent scattering cross section of $B^{(1)}$'s off of quarks q proceeds dominantly through SM Higgs exchange and sub-dominantly (at least for large enough mass splittings $\Delta \equiv (m_{q^{(1)}} - m_{B^{(1)}})/m_{B^{(1)}} \gtrsim 0.1$) through $q^{(1)}$ exchanges.

> **Exercise 223.** Argue that the $B^{(1)}$-quark elastic spin-independent cross section through Higgs exchange goes as m_h^{-2} and that the contribution from $q^{(1)}$ exchange scales as Δ^{-2}.

The spin-dependent cross section proceeds instead exclusively through KK quark exchange, and also scales as Δ^{-2}. Experimentally, spin-independent searches are much more competitive than spin-dependent searches. The latter, however and interestingly, are strongly correlated with the kinetic decoupling temperature (see Ref. [70] and the discussion in Sec. 3.6). Again, we refer the Reader to the thorough review in Ref. [64] for additional details and for a discussion of collider searches (see also Sec. 6.2). It is somewhat heart-warming to know that generation-2 direct detection experiments and the Large Hadron Collider (LHC) will conclusively explore the whole parameter space of at least the minimal incarnation of UEDs, as Fig. A.1, right, shows.

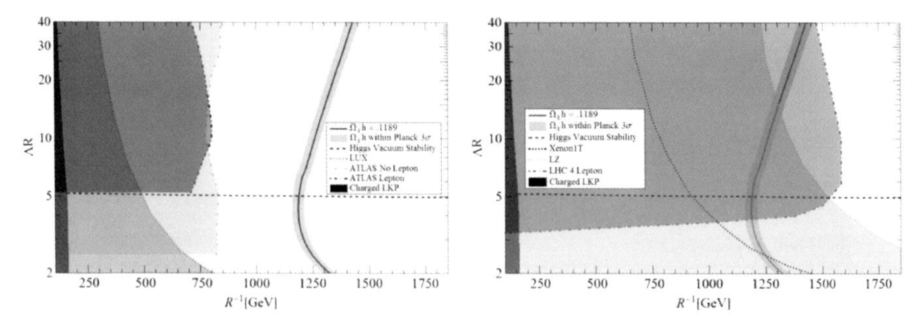

Figure A.1: Current (left) and future (right) collider, direct detection, Higgs vacuum stability, and cosmological limits on the minimal UED parameter space. From Ref. [153].

Appendix B

Rudimentary Cosmology

B.1 Friedman–Robertson–Walker cosmology

The Universe as a whole is (thought to be) a simple object: it is *observed* to be, to a good approximation, spatially isotropic and homogeneous, on large-enough scales.[a] The most general metric on space-time with isotropic and homogeneous spatial sections is known as the Robertson–Walker (RW) metric, which, with no loss of generality, can be cast as (see e.g. [291])

$$ds^2 = dt^2 - R^2(t)\left[\frac{dr^2}{1-kr^2} + r^2 d\Omega\right].$$

Note that the RW metric is invariant under the transformation

$$R \to \lambda^{-1} R,$$
$$r \to \lambda r,$$
$$k \to \lambda^{-2} k,$$

implying that we can choose a normalization where the curvature k is normalized to +1 ("closed", or positive-curvature), 0 (flat) or −1 ("open", in the slang of cosmologists, negative curvature). The *scale factor* $R(t)$ is customarily normalized to its value today, R_0, by defining a dimensionless scale factor $a(t) = R(t)/R_0$. The dynamics of $a(t)$, and thus the RW metric at any time, is dictated by the fundamental equations of General Relativity,

[a]...where the yardstick for "large enough" is a Hubble volume, i.e. roughly a 13 Giga-light-year radius sphere.

Einstein's equations,

$$R_{\mu\nu} = 8\pi G_N \left(T_{\mu\nu} - \frac{1}{2} T g_{\mu\nu} \right),$$

the symbol having the usual meaning: $R_{\mu\nu}$ is Ricci's tensor, $T_{\mu\nu}$ is the energy-momentum tensor, with trace T, and $g_{\mu\nu}$ is the metric tensor. A straightforward and only slightly tedious general relativity exercise gives the non-zero components of the Ricci tensor for the RW metric.

The right-hand side of Einstein's equations calls for a choice for the energy-momentum tensor. In cosmology, matter and energy are customarily modeled as *perfect fluids*, i.e.

$$T^{\mu}_{\nu} = \mathrm{diag}(- \rho, p, p, p),$$

where ρ is the fluid's energy density, and p its pressure, and with trace $T^{\mu}_{\mu} = -\rho + 3p$. Fluids in cosmology are usually assumed to obey some specific equation of state

$$p = w\rho.$$

Conservation of energy implies that

$$\frac{\dot{\rho}}{\rho} = -3(1+w)\frac{\dot{a}}{a}, \tag{B.1}$$

thus, for constant w,

$$\rho \propto a^{-3(1+w)}. \tag{B.2}$$

Exercise 224. Demonstrate Eq. (B.1) using Noether's theorem.

Two key fluids in cosmology are non-relativistic matter ($w = 0$, pressureless fluid), for which Eq. (B.2) gives $\rho_M \sim a^{-3}$, and relativistic matter, or radiation, for which $p_R = \frac{1}{3}\rho_R$, and thus $\rho_R \sim a^{-4}$.

Exercise 225. Demonstrate the equation of state for radiation by calculating the trace of the energy-momentum tensor for the electromagnetic field, and using the equation for the trace of the energy-momentum T.

Finally, vacuum energy can also be considered as a perfect fluid, of equation of state $p_\Lambda = -\rho_\Lambda$, implying $\rho_\Lambda \sim$ const.

Exercise 226. Calculate the equation of state for a spatially constant real scalar field ϕ as a function of the ratio between its kinetic and potential energy (see the discussion in Sec. 3.4).

Now, going back to Einstein's equations, the conscientious Reader will readily verify (Exercise 227) that the $\mu\nu = 00$ equation gives, for a generic perfect fluid,

$$\frac{\ddot{a}}{a} = -4\pi G_N(\rho + 3p), \tag{B.3}$$

while the space-like indices give the three-fold identical (because of isotropy) equation

$$\frac{\ddot{a}}{a} + 2\left(\frac{\dot{a}}{a}\right)^2 + 2\frac{k}{a^2} = 4\pi G_N(\rho - p). \tag{B.4}$$

Upon substitution of Eq. (B.3) into Eq. (B.4) the sedulous Reader will also verify that

$$\left(\frac{\dot{a}}{a}\right)^2 = \frac{8\pi G_N}{3}\rho - \frac{k}{a^2}. \tag{B.5}$$

The left-hand side of Eq. (B.5) is often written as $H^2 \equiv \left(\frac{\dot{a}}{a}\right)^2$, the *Hubble parameter* H. Eqs. (B.3) and (B.5) are known as Friedmann equations. A useful quantity is the critical density

$$\rho_{\text{crit}} \equiv \frac{3H^2}{8\pi G_N}.$$

The ratio of the density of species i to the critical density is indicated with the symbol $\Omega_i \equiv \rho_i/\rho_{\text{crit}}$. The reason why the critical density is in fact critical is that Eq. (B.5) can be recast as

$$\Omega - 1 = \frac{k}{H^2a^2},$$

thus the sign of k, and the (open, closed or flat) "geometry" of the universe, is determined by whether Ω is less, equal or greater than one. Measurements of the cosmic microwave background (CMB) indicate that $\Omega \sim 1$ and that the universe is observationally consistent with being flat. Upon integration of Eq. (B.5), for a universe dominated by an energy density with a power-law dependence on scale factor $\rho \sim a^{-n}$, one gets that

$$a \propto t^{2/n} \qquad (\rho \propto a^{-n}).$$

B.2 Thermodynamics of the early universe

In the dense and hot early universe particles were, presumably, in thermal equilibrium. It is thus useful to use the language and theoretical tools of statistical mechanics, for particles of mass m with g internal degrees of freedom and phase-space distribution function $f(\vec{k})$ (Fermi–Dirac or Bose–Einstein; I am using k to indicate momenta rather than p to avoid confusion with pressure). The number density n, energy density ρ and pressure p are simple integrals of the distribution function:

$$n = \frac{g}{(2\pi)^3} \int f(\vec{k}) d^3k,$$

$$\rho = \frac{g}{(2\pi)^3} \int E(\vec{k}) f(\vec{k}) d^3k,$$

$$p = \frac{g}{(2\pi)^3} \int \frac{|\vec{k}|^2}{3E(\vec{k})} f(\vec{k}) d^3k,$$

with the appropriate relativistic dispersion relation $E^2 = |\vec{k}|^2 + m^2$. The phase-space distribution of a species in *kinetic* equilibrium is the usual

$$f(\vec{k}) = \frac{1}{\exp\left((E-\mu)/T\right) \pm 1},$$

with $+1$ for fermions and -1 for bosons, and μ the chemical potential. If the species is, additionally, in *chemical* equilibrium, then the chemical potentials of species participating in given reactions are correlated. For instance, for a two-particle to two-particle process of the type

$$1 + 2 \leftrightarrow 3 + 4$$

in (chemical) equilibrium, we will have

$$\mu_1 + \mu_2 = \mu_3 + \mu_4.$$

The expressions given above simplify to single integrals in energy as follows:

$$n = \frac{g}{2\pi^2} \int_m^\infty \frac{(E^2 - m^2)^{1/2}}{\exp[(E-\mu)/T] \pm 1} E dE,$$

$$\rho = \frac{g}{2\pi^2} \int_m^\infty \frac{(E^2 - m^2)^{1/2}}{\exp[(E-\mu)/T] \pm 1} E^2 dE,$$

$$p = \frac{g}{6\pi^2} \int_m^\infty \frac{(E^2 - m^2)^{3/2}}{\exp[(E-\mu)/T] \pm 1} dE.$$

It is instructive to derive the relativistic and non-relativistic limits for those expressions. In the $T \gg \mu$ limit (which is most relevant for what we discuss in this book), and in the relativistic limits $T \gg m$, it follows that

$$n = (\zeta(3)/\pi^2) g T^3 \text{ (bosons)}; \quad (3/4)(\zeta(3)/\pi^2) g T^3 \text{ (fermions)} \tag{B.6}$$

$$\rho = (\pi^2/30) g T^4 \text{ (bosons)}; \quad (7/8)(\pi^2/30) g T^4 \text{ (fermions)} \tag{B.7}$$

$$p = \rho/3, \tag{B.8}$$

where $\zeta(3) \approx 1.2\ldots$ is the Riemann zeta function of 3.

Exercise 228. Prove the three Eqs. (B.6)–(B.8) from the corresponding limits of the defining integrals.

The other useful limit is the non-relativistic regime, $T \ll m$, for which

$$n = g\left(\frac{mT}{2\pi}\right)^{3/2} \exp[-(m-\mu)/T], \tag{B.9}$$

$$\rho = m \cdot n, \tag{B.10}$$

$$p = n \cdot T \ll \rho. \tag{B.11}$$

Exercise 229. Ditto, prove the three Eqs. (B.9)–(B.11)

Exercise 230. Show that:

(i) $\langle E \rangle = \frac{\pi^4}{30\zeta(3)} T \simeq 2.7 \cdot T$ (bosons, $\mu = 0$, $T \gg m$),
(ii) $\langle E \rangle = \frac{7\pi^4}{180\zeta(3)} T \simeq 3.15 \cdot T$ (fermions, $\mu = 0$, $T \gg m$),
(iii) $\langle E \rangle = m + \frac{3}{2} T$ ($T \ll m$).

It is convenient to define an effective number of relativistic degrees of freedom g_* as a function of temperature that counts the energy density ρ_R of all relativistic species in the thermal bath (which, if present, dominate the exponentially suppressed energy density of non-relativistic species in thermal equilibrium, at large enough temperature),

$$\rho_R = \frac{\pi^2}{30} g_*(T) T^4,$$

$$g_*(T) \equiv \sum_{\text{bosons } i} g_i \left(\frac{T_i}{T}\right)^4 + \frac{7}{8} \sum_{\text{fermions } i} g_i \left(\frac{T_i}{T}\right)^4.$$

The expression above accounts for the possibility that species i have thermal distributions with a different temperature T_i from that of photons, T. In the very low temperature regime, in the Standard Model

$$g_*(T \ll 1 \text{ MeV}) \simeq 3.36,$$

while at large temperatures

$$g_*(T \gg 300 \text{ GeV}) \gtrsim 106.75.$$

In the early radiation-dominated era ($t \lesssim 4 \times 10^{10}$ s), Friedmann's equation gives [13]

$$H \simeq 1.66 g_*^{1/2}(T) \frac{T^2}{M_P}$$

and

$$t \simeq 0.3 g_*^{-1/2}(T) \frac{M_P}{T^2} \sim \left(\frac{T}{1 \text{ MeV}}\right)^{-2} \text{ s}.$$

A very useful notion in cosmology (and outside cosmology) is that of entropy. In the early universe, entropy per co-moving volume is conserved as long as thermal equilibrium holds. Let us write the second law of thermodynamics for a unit co-moving volume (i.e. for a physical volume $V = R^3$) [13]:

$$TdS = d(\rho V) + pdV = d[(\rho + p)V] - Vdp.$$

Note that the integrability condition

$$\frac{\partial^2 S}{\partial T \partial V} = \frac{\partial^2 S}{\partial V \partial T},$$

gives immediately

$$T\frac{dp}{dT} = \rho + p \rightarrow dp = \frac{\rho + p}{T} dT. \tag{B.12}$$

But then

$$dS = d\left(\frac{(\rho + p)V}{T} + \text{const}\right),$$

Exercise 231. Demonstrate this last equation.

i.e. up to a constant, the entropy per co-moving volume is $S = a^3(\rho + p)/T$. Now let us turn to the first law of thermodynamics;

$$d[(\rho + p)V] = Vdp.$$

Using Eq. (B.12), the equation above gives

$$d\left(\frac{(\rho + p)V}{T}\right) = 0,$$

which means that the entropy per co-moving volume is conserved. Defining the entropy density

$$s \equiv \frac{S}{V} = \frac{\rho + p}{T},$$

since again relativistic species will dominate, at early times, the energy density and pressure of species in thermal equilibrium, we can introduce the effective number of *entropic* relativistic degrees of freedom as a function of temperature

$$g_{*s}(T) \equiv \sum_{\text{bosons } i} g_i \left(\frac{T_i}{T}\right)^3 + \frac{7}{8} \sum_{\text{fermions } i} g_i \left(\frac{T_i}{T}\right)^3,$$

so that

$$s = \frac{2\pi^2}{45} g_{*s}(T) T^3.$$

Since S is conserved $s \sim a^{-3}$ so $g_{*s} T^3 a^3$ ~const. Thus, instead of dividing by a^3 to turn physical to co-moving densities, we can simply divide by s: The relation between the actual number of particles of a given species in thermal equilibrium $N = a^3 n$ is simply the number density of that species divided by s:

$$N = n/s.$$

If number-changing processes are suppressed, i.e. if particles of a given species cannot be created or destroyed, then $n/s \sim$ constant. Another useful fact that stems from conservation of entropy is that $T \propto g_{*s}^{-1/3}(T)/a$: thus, if g_{*s} does not change, $T \sim 1/a$.

B.3 Thermal decoupling

How fast should a reaction be to be in thermal equilibrium? From the discussion above we can derive a good rule of thumb: temperature changes in the early universe by a rate set by the expansion (Hubble) rate, since

$$\frac{\dot{T}}{T} = -\frac{\dot{a}}{a} = -H,$$

so a reaction occurs rapidly enough to maintain thermal distribution as long as its *interaction rate per particle* $\Gamma \equiv n\sigma|v|$ (the product of the reaction cross section σ times the appropriately averaged relative species velocity $|v|$, times number density n) is larger than the Hubble rate,

$$\Gamma \gtrsim H \quad \text{(thermal equilibrium)}.$$

There is an easy way to convince yourself that this criterion works by simply counting the number of reactions expected over the history of the universe, under a few simplifying assumptions. Take $\Gamma \sim T^n$, a power-law in temperature (this book is full of examples of rates with a certain temperature dependence), and assume radiation domination. The number of interactions $N_{\rm int}$ from a given time t on is

$$N_{\rm int} = \int_t^\infty \Gamma(t')dt' = \frac{1}{n-2}\left(\frac{\Gamma}{H}\right)_t.$$

Exercise 232. Prove the equation above.

For $n > 2$ a particle interacts less than one time (i.e. it is effectively decoupled!) after $\Gamma \sim H$.

There is a slightly more formal way to get to the same conclusion. Boltzmann's equation (see Chapter 3) describing the evolution of the number density n of a given particle species can be cast as

$$\frac{dn}{dt} + 3Hn = -\langle\sigma v\rangle(n^2 - n_{\rm eq}^2)$$

(again, see Chapter 3 for the derivation of this equation and for the definition of thermal average $\langle\ldots\rangle$ and of relative velocity v) Defining $Y = n/s$, and $x = m/T$, the equation for the time-evolution of Y reads

$$\frac{x}{Y_{\rm eq}}\frac{dY}{dx} = -\frac{\Gamma}{H}\left[\left(\frac{Y}{Y_{\rm eq}}\right)^2 - 1\right], \quad \Gamma \equiv n_{\rm eq}\langle\sigma v\rangle$$

Exercise 233. Derive this last equation.

In this form, it is clear that what controls the deviation of Y from its equilibrium distribution $Y_{\rm eq}$ is the ratio (Γ/H): for $(\Gamma/H) \ll 1$ Y stops changing, and the number of particles per co-moving volume is thus "locked in". The process of pair-annihilation, that is, *freezes out*. In the opposite regime instead, $(\Gamma/H) \gg 1$, Y is "pushed" dynamically to its equilibrium value by the right-hand side, and the species is in thermal equilibrium. Thermal decoupling, thus, generically occurs around the temperature at which $\Gamma \sim H$.

Bibliography

[1] R. H. Sanders, *The Dark Matter Problem*. UK: Cambridge University Press, Cambridge, 2014.

[2] F. Zwicky, "On the masses of nebulae and of clusters of nebulae," *ApJ* **86** (1937) 217.

[3] Planck Collaboration, P. A. R. Ade *et al.*, "Planck 2015 results. XIII. Cosmological parameters," *Astron. Astrophys.* **594** (2016) A13.

[4] A. G. Riess *et al.*, "A 2.4% Determination of the local value of the Hubble constant," *ApJ* **826** (2016) 56.

[5] J. Binney and S. Tremaine, *Galactic Dynamics: Second Edition*. Princeton University Press, New Jersey, 2008.

[6] K. C. Freeman, "On the disks of spiral and so galaxies," *ApJ* **160** (1970) 811.

[7] F. Hohl, "Dynamical evolution of disk galaxies," *NASA Tech. Rep.* **343** (1970) NASA-TR-R-343, 2+108.

[8] J. P. Ostriker and P. J. E. Peebles, "A numerical study of the stability of flattened galaxies: or, can cold galaxies survive?," *ApJ* **186** (1973) 467–480.

[9] A. Toomre, "What amplifies the spirals," in *Structure and Evolution of Normal Galaxies*, S. M. Fall and D. Lynden-Bell, eds., 1981, pp. 111–136.

[10] F. D'Eramo *et al.*, "Dark matter inelastic up-scattering with the interstellar plasma: An exciting new source of X-Ray lines, including at 3.5 keV," *Phys. Rev.* **D93** no. 10 (2016) 103011.

[11] S. W. Randall *et al.*, "Constraints on the self-interaction cross-section of dark matter from numerical simulations of the merging galaxy cluster 1E 0657-56," *ApJ* **679** (2008) 1173–1180.

[12] S. Dodelson, *Modern Cosmology*. Academic Press, Amsterdam, 2003.

[13] E. W. Kolb and M. S. Turner, *The Early Universe*. 1990.

[14] J. M. O'Meara *et al.*, "The Deuterium to hydrogen abundance ratio towards a fourth QSO: HS 0105 + 1619," *ApJ* **552** (2001) 718–730.

[15] N. A. Bahcall *et al.*, "The mass-to-light function: antibias and Ω_m," *ApJ* **541** (2000) 1–9.

[16] S. E. A. Dodelson, "The three-dimensional power spectrum from angular clustering of galaxies in early sloan digital sky survey data," *ApJ* **572** (2002) 140–156.

[17] M. A. Strauss and J. A. Willick, "The density and peculiar velocity fields of nearby galaxies," *Phys. Rep.* **261** (1995) 271–431.
[18] S. Dodelson, "The real problem with MOND," *Int. J. Mod. Phys.* **D20** (2011) 2749–2753.
[19] W. J. Percival *et al.*, "The shape of the SDSS DR5 galaxy power spectrum," *ApJ* **657** (2007) 645–663.
[20] J. Bekenstein and M. Milgrom, "Does the missing mass problem signal the breakdown of Newtonian gravity?," *ApJ* **286** (1984) 7–14.
[21] M. Zemp *et al.*, "The graininess of dark matter haloes," *Mon. Not. Roy. Astron. Soc.* **394** (2009) 641–659.
[22] M. Lisanti and D. N. Spergel, "Dark matter debris flows in the Milky Way," *Phys. Dark Univ.* **1** (2012) 155–161.
[23] C. W. Purcell, J. S. Bullock, and M. Kaplinghat, "The dark disk of the Milky Way," *ApJ* **703** (2009) 2275–2284.
[24] P. J. McMillan and J. J. Binney, "The uncertainty in Galactic parameters," *Mon. Not. Roy. Astron. Soc.* **402** (2010) 934.
[25] M. C. Smith *et al.*, "The RAVE survey: constraining the local galactic escape speed," *Mon. Not. Roy. Astron. Soc.* **379** (2007) 755–772.
[26] M. Kuhlen *et al.*, "Dark matter direct detection with non-maxwellian velocity structure," *JCAP* **1002** (2010) 030.
[27] B. Paczynski, "Gravitational microlensing by the galactic halo," *ApJ* **304** (1986) 1–5.
[28] L. Wyrzykowski *et al.*, "The OGLE view of microlensing towards the Magellanic Clouds - IV. OGLE-III SMC data and final conclusions on MACHOs," *MNRAS* **416** (2011) 2949–2961.
[29] P. Tisserand *et al.*, "Limits on the Macho content of the Galactic halo from the EROS-2 survey of the magellanic clouds," *A&A* **469** (2007) 387–404.
[30] G. Zaharijas, *Alternative approaches to dark matter puzzle*. PhD thesis, New York U., 2005. Available at: http://wwwlib.umi.com/dissertations/fullcit?p3195499.
[31] K. Jedamzik, "Primordial black hole formation during the QCD epoch," *Phys. Rev. D* **55** (1997) R5871–R5875.
[32] P. H. Frampton *et al.*, "Primordial black holes as all dark matter," *J. Cosmology Astropart. Phys.* **4** (2010) 023.
[33] B. J. Carr *et al.*, "New cosmological constraints on primordial black holes," *Phys. Rev. D* **81** no. 10 (2010) 104019.
[34] M. Y. Khlopov, "Primordial black holes," *Res. Astron. Astrophys.* **10** (2010) 495–528.
[35] A. Barnacka, J.-F. Glicenstein, and R. Moderski, "New constraints on primordial black holes abundance from femtolensing of gamma-ray bursts," *Phys. Rev. D* **86** no. 4 (2012) 043001.
[36] M. Ricotti, J. P. Ostriker, and K. J. Mack, "Effect of primordial black holes on the cosmic microwave background and cosmological parameter estimates," *ApJ* **680** (2008) 829–845.
[37] Virgo, LIGO Scientific Collaboration, B. P. Abbott *et al.*, "Observation of gravitational waves from a binary black hole merger," *Phys. Rev. Lett.* **116** no. 6 (2016) 061102.

[38] S. Bird *et al.*, "Did LIGO detect dark matter?," *Phys. Rev. Lett.* **116** no. 20 (2016) 201301.

[39] J. M. Overduin and P. S. Wesson, "Dark matter and background light," *Phys. Rept.* **402** (2004) 267–406.

[40] K. Sigurdson *et al.*, "Dark-matter electric and magnetic dipole moments," *Phys. Rev.* **D70** (2004) 083501.

[41] M. Pospelov, A. Ritz, and M. B. Voloshin, "Secluded WIMP dark matter," *Phys. Lett.* **B662** (2008) 53–61.

[42] J. Fan *et al.*, "Double-disk dark matter," *Phys. Dark Univ.* **2** (2013) 139–156.

[43] S. Tremaine and J. E. Gunn, "Dynamical role of light neutral leptons in cosmology," *Phys. Rev. Lett.* **42** (1979) 407–410.

[44] C. G. Lacey and J. P. Ostriker, "Massive black holes in galactic halos?," *ApJ* **299** (1985) 633–652.

[45] T. Goerdt *et al.*, "The survival and disruption of CDM micro-haloes: Implications for direct and indirect detection experiments," *Mon. Not. Roy. Astron. Soc.* **375** (2007) 191–198.

[46] N. Afshordi, P. McDonald, and D. N. Spergel, "Primordial black holes as dark matter: The power spectrum and evaporation of early structures," *ApJ* **594** (2003) L71–L74.

[47] S. Weinberg, *Gravitation and Cosmology: Principles and Applications of the General Theory of Relativity*. 1972.

[48] R. Cowsik and J. McClelland, "An upper limit on the neutrino rest mass," *Phys. Rev. Lett.* **29** (1972) 669–670.

[49] C. Boehm, M. J. Dolan, and C. McCabe, "A lower bound on the mass of cold thermal dark matter from planck," *JCAP* **1308** (2013) 041.

[50] J. L. Feng and J. Kumar, "The WIMPless miracle: Dark-matter particles without weak-scale masses or weak interactions," *Phys. Rev. Lett.* **101** (2008) 231301.

[51] K. Griest and M. Kamionkowski, "Unitarity limits on the mass and radius of dark matter particles," *Phys. Rev. Lett.* **64** (1990) 615.

[52] B. W. Lee and S. Weinberg, "Cosmological lower bound on heavy neutrino masses," *Phys. Rev. Lett.* **39** (1977) 165–168.

[53] S. Profumo, "Hunting the lightest lightest neutralinos," *Phys. Rev.* **D78** (2008) 023507.

[54] E. A. Baltz, "Dark matter candidates," *eConf* **C040802** (2004) L002.

[55] C. Wainwright and S. Profumo, "The Impact of a strongly first-order phase transition on the abundance of thermal relics," *Phys. Rev.* **D80** (2009) 103517.

[56] K. Griest and D. Seckel, "Three exceptions in the calculation of relic abundances," *Phys. Rev.* **D43** (1991) 3191–3203.

[57] P. Gondolo and G. Gelmini, "Cosmic abundances of stable particles: Improved analysis," *Nucl. Phys.* **B360** (1991) 145–179.

[58] J. Edsjo *et al.*, "Accurate relic densities with neutralino, chargino and sfermion coannihilations in mSUGRA," *JCAP* **0304** (2003) 001.

[59] P. Gondolo *et al.*, "DarkSUSY: Computing supersymmetric dark matter properties numerically," *JCAP* **0407** (2004) 008.

[60] G. Belanger *et al.*, "MicrOMEGAs: A program for calculating the relic density in the MSSM," *Comput. Phys. Commun.* **149** (2002) 103–120.
[61] S. Profumo and P. Ullio, "SUSY dark matter and quintessence," *JCAP* **0311** (2003) 006.
[62] R. T. D'Agnolo and J. T. Ruderman, "Light dark matter from forbidden channels," *Phys. Rev. Lett.* **115** no. 6 (2015) 061301.
[63] J. Edsjo and P. Gondolo, "Neutralino relic density including coannihilations," *Phys. Rev.* **D56** (1997) 1879–1894.
[64] D. Hooper and S. Profumo, "Dark matter and collider phenomenology of universal extra dimensions," *Phys. Rept.* **453** (2007) 29–115.
[65] M. Citron *et al.*, "End of the CMSSM coannihilation strip is nigh," *Phys. Rev.* **D87** no. 3 (2013) 036012.
[66] P. Salati, "Quintessence and the relic density of neutralinos," *Phys. Lett.* **B571** (2003) 121–131.
[67] E. Kuflik *et al.*, "Elastically decoupling dark matter".
[68] A. M. Green, S. Hofmann, and D. J. Schwarz, "The power spectrum of SUSY-CDM on subgalactic scales," *MNRAS* **353** (2004) L23–L27.
[69] A. M. Green, S. Hofmann, and D. J. Schwarz, "The first WIMPy halos," *J. Cosmology Astropart. Phys.* **8** (2005) 003.
[70] J. M. Cornell and S. Profumo, "Earthly probes of the smallest dark matter halos," *JCAP* **1206** (2012) 011.
[71] A. Loeb and M. Zaldarriaga, "Small-scale power spectrum of cold dark matter," *Phys. Rev. D* **71** no. 10 (May, 2005) 103520.
[72] E. Bertschinger, "Effects of cold dark matter decoupling and pair annihilation on cosmological perturbations," *Phys. Rev. D* **74** no. 6 (2006) 063509.
[73] S. Profumo, K. Sigurdson, and M. Kamionkowski, "What mass are the smallest protohalos?," *Phys. Rev. Lett.* **97** (2006) 031301.
[74] E. D. Carlson, M. E. Machacek, and L. J. Hall, "Self-interacting dark matter," *ApJ* **398** (1992) 43–52.
[75] W. J. G. de Blok, "The Core-Cusp Problem," *Adv. Astron.* **2010** (2010) 789293.
[76] M. Boylan-Kolchin, J. S. Bullock, and M. Kaplinghat, "Too big to fail? The puzzling darkness of massive Milky Way subhaloes," *Mon. Not. Roy. Astron. Soc.* **415** (2011) L40.
[77] Y. Hochberg *et al.*, "Mechanism for thermal relic dark matter of strongly interacting massive particles," *Phys. Rev. Lett.* **113** (2014) 171301.
[78] F. D'Eramo and J. Thaler, "Semi-annihilation of Dark Matter," *JHEP* **06** (2010) 109.
[79] F. D'Eramo, M. McCullough, and J. Thaler, "Multiple gamma lines from semi-annihilation," *JCAP* **1304** (2013) 030.
[80] L. J. Hall *et al.*, "Freeze-in production of FIMP dark matter," *JHEP* **03** (2010) 080.
[81] R. T. Co *et al.*, "Freeze-in dark matter with displaced signatures at colliders," *JCAP* **1512** no. 12 (2015) 024.
[82] R. Peierls and H. A. Bethe, "The neutrino," *Nature* **133** (1934) 532.
[83] A. Franklin, *Shifting Standards: Experiments in Particle Physics in the Twentieth Century*. U. Pittsburgh Press, 2014.

[84] F. Reines and C. L. Cowan, "Detection of the free neutrino," *Phys. Rev.* **92** (1953) 830–831.

[85] C. L. Cowan *et al.*, "Detection of the free neutrino: A confirmation," *Science* **124** (1956) 103–104.

[86] G. Barello, S. Chang, and C. A. Newby, "A model independent approach to inelastic dark matter scattering," *Phys. Rev.* **D90** no. 9 (2014) 094027.

[87] D. Tucker-Smith and N. Weiner, "Inelastic dark matter," *Phys. Rev.* **D64** (2001) 043502.

[88] Gargamelle Neutrino Collaboration, F. J. Hasert *et al.*, "Observation of neutrino like interactions without muon or electron in the gargamelle neutrino experiment," *Phys. Lett.* **B46** (1973) 138–140.

[89] M. W. Goodman and E. Witten, "Detectability of certain dark matter candidates," *Phys. Rev.* **D31** (1985) 3059.

[90] R. Essig, J. Mardon, and T. Volansky, "Direct detection of sub-GeV dark matter," *Phys. Rev.* **D85** (2012) 076007.

[91] P. W. Graham *et al.*, "Semiconductor probes of light dark matter," *Phys. Dark Univ.* **1** (2012) 32–49.

[92] Y. Hochberg, Y. Zhao, and K. M. Zurek, "Superconducting detectors for superlight dark matter," *Phys. Rev. Lett.* **116** no. 1 (2016) 011301.

[93] A. Drukier and L. Stodolsky, "Principles and applications of a neutral current detector for neutrino physics and astronomy," *Phys. Rev.* **D30** (1984) 2295.

[94] A. K. Drukier, K. Freese, and D. N. Spergel, "Detecting cold dark matter candidates," *Phys. Rev.* **D33** (1986) 3495–3508.

[95] J. I. Collar and F. T. Avignone, III, "The effect of elastic scattering in the earth on cold dark matter experiments," *Phys. Rev.* **D47** (1993) 5238–5246.

[96] D. N. Spergel, "The motion of the earth and the detection of Wimps," *Phys. Rev.* **D37** (1988) 1353.

[97] A. L. Fitzpatrick *et al.*, "The effective field theory of dark matter direct detection," *JCAP* **1302** (2013) 004.

[98] P. Gondolo, "Phenomenological introduction to direct dark matter detection," in *Dark matter in cosmology, quantum measurements, experimental gravitation. Proceedings, 31st Rencontres de Moriond, 16th Moriond Workshop*, Les Arcs, France, January 2–27, 1996, pp. 41–51.

[99] J. Kumar and D. Marfatia, "Matrix element analyses of dark matter scattering and annihilation," *Phys. Rev.* **D88** no. 1 (2013) 014035.

[100] G. Jungman, M. Kamionkowski, and K. Griest, "Supersymmetric dark matter," *Phys. Rept.* **267** (1996) 195–373.

[101] J. R. Ellis and R. A. Flores, "Realistic predictions for the detection of supersymmetric dark matter," *Nucl. Phys.* **B307** (1988) 883.

[102] J. Engel and P. Vogel, "Spin dependent cross-sections of weakly interacting massive particles on nuclei," *Phys. Rev.* **D40** (1989) 3132–3135.

[103] M. T. Ressell *et al.*, "Nuclear shell model calculations of neutralino — nucleus cross-sections for Si-29 and Ge-73," *Phys. Rev.* **D48** (1993) 5519–5535.

[104] J. Cooley, "Overview of non-liquid noble direct detection dark matter experiments," *Phys. Dark Univ.* **4** (2014) 92–97.

[105] F. D'Eramo and M. Procura, "Connecting dark matter UV complete models to direct detection rates via effective field theory," *JHEP* **04** (2015) 054.

[106] M. A. Shifman, A. I. Vainshtein, and V. I. Zakharov, "Remarks on Higgs Boson interactions with nucleons," *Phys. Lett.* **B78** (1978) 443–446.

[107] A. I. Vainshtein, V. I. Zakharov, and M. A. Shifman, "Higgs particles," *Sov. Phys. Usp.* **23** (1980) 429–449.

[108] V. Barger, W.-Y. Keung, and G. Shaughnessy, "Spin dependence of dark matter scattering," *Phys. Rev.* **D78** (2008) 056007.

[109] A. Crivellin, F. D'Eramo, and M. Procura, "New constraints on dark matter effective theories from standard model loops," *Phys. Rev. Lett.* **112** (2014) 191304.

[110] J. R. Primack, D. Seckel, and B. Sadoulet, "Detection of cosmic dark matter," *Ann. Rev. Nucl. Part. Sci.* **38** (1988) 751–807.

[111] J. Monroe and P. Fisher, "Neutrino backgrounds to dark matter searches," *Phys. Rev.* **D76** (2007) 033007.

[112] J. Billard, L. Strigari, and E. Figueroa-Feliciano, "Implication of neutrino backgrounds on the reach of next generation dark matter direct detection experiments," *Phys. Rev.* **D89** no. 2 (2014) 023524.

[113] D. G. Cerdeno *et al.*, "Physics from solar neutrinos in dark matter direct detection experiments," *JHEP* **05** (2016) 118.

[114] S. Profumo and L. Ubaldi, "Cosmic ray-dark matter scattering: a new signature of (asymmetric) dark matter in the gamma ray sky," *JCAP* **1108** (2011) 020.

[115] S. Profumo, L. Ubaldi, and M. Gorchtein, "Gamma rays from cosmic-ray proton scattering in AGN jets: the intra-cluster gas vastly outshines dark matter," *JCAP* **1304** (2013) 012.

[116] M. Gorchtein, S. Profumo, and L. Ubaldi, "Probing dark matter with AGN jets," *Phys. Rev.* **D82** (2010) 083514.

[117] S. Profumo and K. Sigurdson, "The shadow of dark matter," *Phys. Rev.* **D75** (2007) 023521.

[118] Y. Mambrini, S. Profumo, and F. S. Queiroz, "Dark matter and global symmetries," *Phys. Lett.* **B760** (2016) 807–815.

[119] PAMELA Collaboration, O. Adriani *et al.*, "An anomalous positron abundance in cosmic rays with energies 1.5-100 GeV," *Nature* **458** (2009) 607–609.

[120] AMS Collaboration, M. Aguilar *et al.*, "First result from the alpha magnetic spectrometer on the international space station: Precision measurement of the positron fraction in primary cosmic rays of 0.5-350 GeV," *Phys. Rev. Lett.* **110** (2013) 141102.

[121] A. Cooray, "Extragalactic background light measurements and applications," *Royal Society Open Science* **3** (2016) 150555, doi: 10.1098/rsos.150555.

[122] N. Arkani-Hamed, A. Delgado, and G. F. Giudice, "The Well-tempered neutralino," *Nucl. Phys.* **B741** (2006) 108–130.

[123] M. Cirelli, A. Strumia, and M. Tamburini, "Cosmology and astrophysics of minimal dark matter," *Nucl. Phys.* **B787** (2007) 152–175.

[124] S. D. McDermott, H.-B. Yu, and K. M. Zurek, "Constraints on scalar asymmetric dark matter from black hole formation in neutron stars," *Phys. Rev.* **D85** (2012) 023519.
[125] G. G. Raffelt, *Stars as Laboratories for Fundamental Physics*. USA: University Press, Chicago, 1996, 664 p.
[126] D. Spolyar *et al.*, "Dark stars: Dod och ateruppstandelse," *PoS* **IDM2008** (2008) 077.
[127] L. M. Krauss, M. Srednicki, and F. Wilczek, "Solar system constraints and signatures for dark matter candidates," *Phys. Rev.* **D33** (1986) 2079–2083.
[128] G. D. Mack, J. F. Beacom, and G. Bertone, "Towards closing the window on strongly interacting dark matter: Far-reaching constraints from earth's heat flow," *Phys. Rev.* **D76** (2007) 043523.
[129] M. Cirelli *et al.*, "PPPC 4 DM ID: A poor particle physicist cookbook for dark matter indirect detection," *JCAP* **1103** (2011) 051.
[130] PAMELA Collaboration, O. Adriani *et al.*, "PAMELA results on the cosmic-ray antiproton flux from 60 MeV to 180 GeV in kinetic energy," *Phys. Rev. Lett.* **105** (2010) 121101.
[131] T. Aramaki *et al.*, "Antideuterons as an indirect dark matter signature: Si(Li) detector development and a GAPS balloon mission," *Adv. Space Res.* **46** (2010) 1349–1353.
[132] Fermi-LAT Collaboration, M. Ackermann *et al.*, "Measurement of separate cosmic-ray electron and positron spectra with the Fermi Large Area Telescope," *Phys. Rev. Lett.* **108** (2012) 011103.
[133] S. Profumo, "Dissecting cosmic-ray electron-positron data with Occam's Razor: the role of known Pulsars," *Central Eur. J. Phys.* **10** (2011) 1–31.
[134] L. Gendelev, S. Profumo, and M. Dormody, "The contribution of fermi gamma-ray pulsars to the local flux of cosmic-ray electrons and positrons," *JCAP* **1002** (2010) 016.
[135] M. S. Longair, ed., *High-Energy Astrophysics. Vol. 1: Particles, Photons and Their Detection*. UK: University Press, Cambridge, 1992, 418 p.
[136] S. Profumo, "An observable electron-positron anisotropy cannot be generated by dark matter," *JCAP* **1502** no. 02 (2015) 043.
[137] T. Linden and S. Profumo, "Probing the pulsar origin of the anomalous positron fraction with AMS-02 and atmospheric cherenkov telescopes," *ApJ* **772** (2013) 18.
[138] IceCube, ANTARES Collaboration, S. Adrian-Martinez *et al.*, "The first combined search for neutrino point-sources in the southern hemisphere with the antares and icecube neutrino telescopes," *ApJ* **823** no. 1 (2016) 65.
[139] J. L. Feng, J. Smolinsky, and P. Tanedo, "Detecting dark matter through dark photons from the Sun: Charged particle signatures," *Phys. Rev.* **D93** no. 11 (2016) 115036.
[140] P. J. Fox and E. Poppitz, "Leptophilic dark matter," *Phys. Rev.* **D79** (2009) 083528.
[141] T. E. Jeltema and S. Profumo, "Dark matter detection with hard x-ray telescopes," *Mon. Not. Roy. Astron. Soc.* **421** (2012) 1215.

[142] S. Colafrancesco, S. Profumo, and P. Ullio, "Multi-frequency analysis of neutralino dark matter annihilations in the coma cluster," *Astron. Astrophys.* **455** (2006) 21.

[143] L. Bergstrom and P. Ullio, "Full one loop calculation of neutralino annihilation into two photons," *Nucl. Phys.* **B504** (1997) 27–44.

[144] Fermi-LAT Collaboration, M. Ackermann *et al.*, "Constraining dark matter models from a combined analysis of milky way satellites with the fermi large area telescope," *Phys. Rev. Lett.* **107** (2011) 241302.

[145] C. Weniger, "A tentative gamma-ray line from dark matter annihilation at the fermi large area telescope," *JCAP* **1208** (2012) 007.

[146] M. Su and D. P. Finkbeiner, "Strong evidence for gamma-ray line emission from the inner galaxy," (2012).

[147] E. Carlson *et al.*, "Clustering analysis of the morphology of the 130 GeV gamma-ray feature," *Phys. Rev.* **D88** no. 4 (2013) 043006.

[148] F. Aharonian, D. Khangulyan, and D. Malyshev, "Cold ultrarelativistic pulsar winds as potential sources of galactic gamma-ray lines above 100 GeV," *Astron. Astrophys.* **547** (2012) A114.

[149] A. Ibarra *et al.*, "Gamma-ray boxes from axion-mediated dark matter," *JCAP* **1305** (2013) 016.

[150] A. Ibarra, S. L. Gehler, and M. Pato, "Dark matter constraints from box-shaped gamma-ray features," *JCAP* **1207** (2012) 043.

[151] H.-C. Cheng, K. T. Matchev, and M. Schmaltz, "Bosonic supersymmetry? Getting fooled at the CERN LHC," *Phys. Rev.* **D66** (2002) 056006.

[152] ATLAS Collaboration, T. A. collaboration, "Search for squarks and gluinos with the ATLAS detector in final states with jets and missing transverse momentum and 20.3 fb^{-1} of $\sqrt{s} = 8$ TeV proton–proton collision data," (2013).

[153] J. M. Cornell, S. Profumo, and W. Shepherd, "Dark matter in minimal universal extra dimensions with a stable vacuum and the "right" Higgs boson," *Phys. Rev.* **D89** no. 5 (2014) 056005.

[154] J. Goodman *et al.*, "Constraints on dark matter from colliders," *Phys. Rev.* **D82** (2010) 116010.

[155] S. Profumo, W. Shepherd, and T. Tait, "Pitfalls of dark matter crossing symmetries," *Phys. Rev.* **D88** no. 5 (2013) 056018.

[156] J. Abdallah *et al.*, "Simplified models for dark matter and missing energy searches at the LHC" (2014).

[157] J. Abdallah *et al.*, "Simplified models for dark matter searches at the LHC," *Phys. Dark Univ.* **9–10** (2015) 8–23.

[158] H. Davoudiasl, R. Kitano, T. Li, and H. Murayama, "The new minimal standard model," *Phys. Lett.* **B609** (2005) 117–123.

[159] R. Mahbubani and L. Senatore, "The minimal model for dark matter and unification," *Phys. Rev.* **D73** (2006) 043510.

[160] F. D'Eramo, "Dark matter and Higgs boson physics," *Phys. Rev.* **D76** (2007) 083522.

[161] ATLAS Collaboration, G. Aad *et al.*, "Search for invisible decays of a Higgs boson using vector-boson fusion in pp collisions at $\sqrt{s} = 8$ TeV with the ATLAS detector," *JHEP* **01** (2016) 172.

[162] G. 't Hooft, "Computation of the quantum effects due to a four-dimensional pseudoparticle," *Phys. Rev.* **D14** (1976) 3432–3450.
[163] S. L. Adler, "Axial vector vertex in spinor electrodynamics," *Phys. Rev.* **177** (1969) 2426–2438.
[164] J. S. Bell and R. Jackiw, "A PCAC puzzle: pi0 –> gamma gamma in the sigma model," *Nuovo Cim.* **A60** (1969) 47–61.
[165] C. Vafa and E. Witten, "Parity Conservation in QCD," *Phys. Rev. Lett.* **53** (1984) 535.
[166] A. A. Anselm and N. G. Uraltsev, "A second massless axion?," *Phys. Lett.* **B114** (1982) 39–41.
[167] A. G. Dias *et al.*, "The quest for an intermediate-scale accidental axion and further ALPs," *JHEP* **06** (2014) 037.
[168] E. Witten, "Some properties of O(32) superstrings," *Phys. Lett.* **B149** (1984) 351–356.
[169] J. P. Conlon, "The QCD axion and moduli stabilisation," *JHEP* **05** (2006) 078.
[170] A. Arvanitaki *et al.*, "String Axiverse," *Phys. Rev.* **D81** (2010) 123530.
[171] P. Sikivie, "Axions," in Bertone, G. (ed.): *Particle Dark Matter*, 2010, pp. 204–227.
[172] G. G. Raffelt, "Astrophysical methods to constrain axions and other novel particle phenomena," *Phys. Rept.* **198** (1990) 1–113.
[173] J. Engel, D. Seckel, and A. C. Hayes, "Emission and detectability of hadronic axions from SN1987A," *Phys. Rev. Lett.* **65** (1990) 960–963.
[174] P. Sikivie, "Axion cosmology," *Lect. Notes Phys.* **741** (2008) 19–50.
[175] S. Hannestadi *et al.*, "Neutrino and axion hot dark matter bounds after WMAP-7," *JCAP* **1008** (2010) 001.
[176] K. J. Bae, J.-H. Huh, and J. E. Kim, "Update of axion CDM energy," *JCAP* **0809** (2008) 005.
[177] M. S. Turner, "Cosmic and Local Mass Density of Invisible Axions," *Phys. Rev.* **D33** (1986) 889–896.
[178] F. D'Eramo, L. J. Hall, and D. Pappadopulo, "Multiverse dark matter: SUSY or axions," *JHEP* **11** (2014) 108.
[179] T. Hiramatsu *et al.*, "Production of dark matter axions from collapse of string-wall systems," *Phys. Rev.* **D85** (2012) 105020.
[180] M. S. Turner, "Axions from SN 1987a," *Phys. Rev. Lett.* **60** (1988) 1797.
[181] D. Grin *et al.*, "A telescope search for decaying relic axions," *Phys. Rev.* **D75** (2007) 105018.
[182] B. D. Blout *et al.*, "A radio telescope search for axions," *ApJ* **546** (2001) 825–828.
[183] H. Murayama *et al.*, "Axions and other very light bosons," *Eur. Phys. J.* **C15** (2000) 298–305.
[184] F. Della Valle *et al.*, "The PVLAS experiment: Measuring vacuum magnetic birefringence and dichroism with a birefringent Fabry-Perot cavity," *Eur. Phys. J.* **C76** no. 1 (2016) 24.
[185] PVLAS Collaboration, E. Zavattini *et al.*, "New PVLAS results and limits on magnetically induced optical rotation and ellipticity in vacuum," *Phys. Rev.* **D77** (2008) 032006.
[186] P. Sikivie, "Experimental tests of the invisible axion," *Phys. Rev. Lett.* **51** (1983) 1415–1417.

[187] G. Carosi *et al.*, "Probing the axion-photon coupling: Phenomenological and experimental perspectives. A snowmass white paper," in *Community Summer Study 2013: Snowmass on the Mississippi (CSS2013) Minneapolis*, MN, USA, July 29–August 6, 2013.

[188] J. Schechter and J. W. F. Valle, "Neutrino masses in SU(2) x U(1) theories," *Phys. Rev.* **D22** (1980) 2227.

[189] G. 't Hooft, "Naturalness, chiral symmetry, and spontaneous chiral symmetry breaking," *NATO Sci. Ser. B* **59** (1980) 135.

[190] S. Dodelson and L. M. Widrow, "Sterile-neutrinos as dark matter," *Phys. Rev. Lett.* **72** (1994) 17–20.

[191] M. Shaposhnikov, "Sterile neutrinos," in In Bertone, G. (ed.): Particle Dark Matter, 2010, pp. 228–248.

[192] A. Kusenko, "Sterile neutrinos: The dark side of the light fermions," *Phys. Rept.* **481** (2009) 1–28.

[193] L. Wolfenstein, "Neutrino oscillations in matter," *Phys. Rev.* **D17** (1978) 2369–2374.

[194] S. P. Mikheev and A. Yu. Smirnov, "Resonance amplification of oscillations in matter and spectroscopy of solar neutrinos," *Sov. J. Nucl. Phys.* **42** (1985) 913–917.

[195] X.-D. Shi and G. M. Fuller, "A new dark matter candidate: Nonthermal sterile neutrinos," *Phys. Rev. Lett.* **82** (1999) 2832–2835.

[196] T. Asaka, S. Blanchet, and M. Shaposhnikov, "The nuMSM, dark matter and neutrino masses," *Phys. Lett.* **B631** (2005) 151–156.

[197] M. Shaposhnikov, "A possible symmetry of the nuMSM," *Nucl. Phys.* **B763** (2007) 49–59.

[198] J. R. Bond, G. Efstathiou, and J. Silk, "Massive neutrinos and the large scale structure of the universe," *Phys. Rev. Lett.* **45** (1980) 1980–1984.

[199] U. Seljak *et al.*, "Can sterile neutrinos be the dark matter?," *Phys. Rev. Lett.* **97** (2006) 191303.

[200] M. Viel *et al.*, "Can sterile neutrinos be ruled out as warm dark matter candidates?," *Phys. Rev. Lett.* **97** (2006) 071301.

[201] M. Viel *et al.*, "How cold is cold dark matter? Small scales constraints from the flux power spectrum of the high-redshift Lyman-alpha forest," *Phys. Rev. Lett.* **100** (2008) 041304.

[202] J. Baur *et al.*, "Lyman-alpha forests cool warm dark matter," in *SDSS-IV Collaboration Meeting*, July 20–23, 2015.

[203] L. C. Loveridge, "Gravitational waves from a pulsar kick caused by neutrino conversions," *Phys. Rev.* **D69** (2004) 024008.

[204] E. Bulbul *et al.*, "Detection of an unidentified emission line in the stacked x-ray spectrum of galaxy clusters," *ApJ* **789** (2014) 13.

[205] A. Boyarsky *et al.*, "Unidentified line in x-ray spectra of the andromeda galaxy and perseus galaxy cluster," *Phys. Rev. Lett.* **113** (2014) 251301.

[206] T. E. Jeltema and S. Profumo, "Discovery of a 3.5 keV line in the galactic centre and a critical look at the origin of the line across astronomical targets," *Mon. Not. Roy. Astron. Soc.* **450** no. 2 (2015) 2143–2152.

[207] T. Jeltema and S. Profumo, "Reply to two comments on "Dark matter searches going bananas the contribution of Potassium (and Chlorine) to the 3.5 keV line"" (2014).

[208] K. J. H. Phillips, B. Sylwester, and J. Sylwester, "The x-ray line feature at 3.5 KeV in galaxy cluster spectra," *ApJ* **809** (2015) 50.

[209] M. E. Anderson, E. Churazov, and J. N. Bregman, "Non-detection of x-ray emission from sterile neutrinos in stacked galaxy spectra," *Mon. Not. Roy. Astron. Soc.* **452** no. 4 (2015) 3905–3923.

[210] D. Malyshev, A. Neronov, and D. Eckert, "Constraints on 3.55 keV line emission from stacked observations of dwarf spheroidal galaxies," *Phys. Rev.* **D90** (2014) 103506.

[211] O. Urban *et al.*, "A suzaku search for dark matter emission lines in the x-ray brightest galaxy clusters," *Mon. Not. Roy. Astron. Soc.* **451** no. 3 (2015) 2447–2461.

[212] T. E. Jeltema and S. Profumo, "Deep XMM observations of Draco rule out at the 99% confidence level a dark matter decay origin for the 3.5 keV line," *Mon. Not. Roy. Astron. Soc.* **458** (2016) 3592.

[213] E. Carlson, T. Jeltema, and S. Profumo, "Where do the 3.5 keV photons come from? A morphological study of the Galactic Center and of Perseus," *JCAP* **1502** no. 02 (2015) 009.

[214] M. Cicoli *et al.*, "3.55 keV photon line and its morphology from a 3.55 keV axionlike particle line," *Phys. Rev.* **D90** (2014) 023540.

[215] G. Bertone and D. Hooper, "A history of dark matter" (2016).

[216] H. Pagels and J. R. Primack, "Supersymmetry, cosmology and new TeV physics," *Phys. Rev. Lett.* **48** (1982) 223.

[217] V. S. Rychkov and A. Strumia, "Thermal production of gravitinos," *Phys. Rev.* **D75** (2007) 075011.

[218] C. Cheung, Y. Nomura, and J. Thaler, "Goldstini," *JHEP* **03** (2010) 073.

[219] C. Cheung, F. D'Eramo, and J. Thaler, "The spectrum of goldstini and modulini," *JHEP* **08** (2011) 115.

[220] G. F. Giudice and R. Rattazzi, "Theories with gauge mediated supersymmetry breaking," *Phys. Rept.* **322** (1999) 419–499.

[221] T. Gherghetta, "Goldstino decoupling in spontaneously broken supergravity theories," *Nucl. Phys.* **B485** (1997) 25–37.

[222] T. Moroi, H. Murayama, and M. Yamaguchi, "Cosmological constraints on the light stable gravitino," *Phys. Lett.* **B303** (1993) 289–294.

[223] M. Drees, R. Godbole, and P. Roy, *Theory and Phenomenology of Sparticles: An Account of Four-Dimensional N=1 Supersymmetry in High Energy Physics*. USA: World Scientific, Hackensack, 2004, 555 p.

[224] S. Bailly, K. Jedamzik, and G. Moultaka, "Gravitino dark matter and the cosmic lithium abundances," *Phys. Rev.* **D80** (2009) 063509.

[225] M. Kaplinghat, "Dark matter from early decays," *Phys. Rev.* **D72** (2005) 063510.

[226] I. F. M. Albuquerque, G. Burdman, and Z. Chacko, "Direct detection of supersymmetric particles in neutrino telescopes," *Phys. Rev.* **D75** (2007) 035006.

[227] M. Drees and X. Tata, "Signals for heavy exotics at hadron colliders and supercolliders," *Phys. Lett.* **B252** (1990) 695–702.

[228] J. L. Feng and T. Moroi, "Tevatron signatures of longlived charged sleptons in gauge mediated supersymmetry breaking models," *Phys. Rev.* **D58** (1998) 035001.

[229] J. L. Feng and B. T. Smith, "Slepton trapping at the large hadron and international linear colliders," *Phys. Rev.* **D71** (2005) 015004.

[230] N. E. Bomark *et al.*, "Photon, neutrino and charged particle spectra from R-violating gravitino decays," *Phys. Lett.* **B686** (2010) 152–161.

[231] E. Schrödinger, "The proper vibrations of the expanding universe," *Physica* **6** (1939) 899–912.

[232] N. A. Chernikov and E. A. Tagirov, "Quantum theory of scalar fields in de Sitter space-time," *Annales Poincare Phys. Theor.* **A9** (1968) 109.

[233] L. Parker, "Particle creation in expanding universes," *Phys. Rev. Lett.* **21** (1968) 562–564.

[234] A. A. Grib and S. G. Mamaev, "On field theory in the friedman space," *Yad. Fiz.* **10** (1969) 1276–1281. [*Sov. J. Nucl. Phys.* 10, (1970) 722].

[235] Ya. B. Zeldovich and A. A. Starobinsky, "Particle production and vacuum polarization in an anisotropic gravitational field," *Sov. Phys. JETP* **34** (1972) 1159–1166. [Zh. Eksp. Teor. Fiz. 61, (1971) 2161].

[236] D. J. H. Chung, E. W. Kolb, and A. Riotto, "Superheavy dark matter," *Phys. Rev.* **D59** (1999) 023501.

[237] E. W. Kolb, D. J. H. Chung, and A. Riotto, "WIMPzillas!," in *Trends in Theoretical Physics II. Proceedings, 2nd La Plata Meeting*, Buenos Aires, Argentina, November 29–December 4, 1998, pp. 91–105.

[238] BICEP2 Collaboration, P. A. R. Ade *et al.*, "Detection of *B*-Mode polarization at degree angular scales by BICEP2," *Phys. Rev. Lett.* **112** no. 24 (2014) 241101.

[239] M. J. Mortonson and U. Seljak, "A joint analysis of Planck and BICEP2 B modes including dust polarization uncertainty," *JCAP* **1410** (2014) 035.

[240] D. J. H. Chung *et al.*, "Isocurvature constraints on gravitationally produced superheavy dark matter," *Phys. Rev.* **D72** (2005) 023511.

[241] V. A. Kuzmin and I. I. Tkachev, "Ultrahigh-energy cosmic rays and inflation relics," *Phys. Rept.* **320** (1999) 199–221.

[242] Y. L. Dokshitzer *et al.*, *Basics of Perturbative QCD*. France: Ed. Frontieres, Gif-sur-Yvette, 1991, 274 p.

[243] P. Blasi, R. Dick, and E. W. Kolb, "Ultrahigh-energy cosmic rays from annihilation of superheavy dark matter," *ApJ* **18** (2002) 57–66.

[244] E. Witten, "Cosmic separation of phases," *Phys. Rev.* **D30** (1984) 272–285.

[245] K. Lawson and A. R. Zhitnitsky, "Quark (Anti) nugget dark matter," in *Cosmic Frontier Workshop: Snowmass 2013 Menlo Park, USA, March 6–8, 2013*.

[246] E. Carlson and S. Profumo, "When dark matter interacts with cosmic rays or interstellar matter: A morphological study," *Phys. Rev.* **D92** no. 6 (2015) 063003.

[247] M. Dine and A. Kusenko, "The Origin of the matter — antimatter asymmetry," *Rev. Mod. Phys.* **76** (2003) 1.

[248] J. R. Primack, "Cosmology: Small scale issues revisited," *New J. Phys.* **11** (2009) 105029.

[249] A. B. Newman *et al.*, "The density profiles of massive, relaxed galaxy clusters: II. separating luminous and dark matter in cluster cores," *ApJ* **765** (2013) 25.

[250] K. K. Boddy *et al.*, "Self-interacting dark matter from a non-abelian hidden sector," *Phys. Rev.* **D89** no. 11 (2014) 115017.

[251] S. Tulin, H.-B. Yu, and K. M. Zurek, "Beyond collisionless dark matter: Particle physics dynamics for dark matter halo structure," *Phys. Rev.* **D87** no. 11 (2013) 115007.

[252] Y. Mambrini and T. Toma, "X-ray lines and self-interacting dark matter," *Eur. Phys. J.* **C75** no. 12 (2015) 570.

[253] S. Nussinov, "Technocosmology — could a technibaryon excess provide a "natural" missing mass candidate?," *Phys. Lett. B* **165** (1985) 55–58.

[254] D. E. Kaplan, M. A. Luty, and K. M. Zurek, "Asymmetric dark matter," *Phys. Rev.* **D79** (2009) 115016.

[255] K. M. Zurek, "Asymmetric dark matter: Theories, signatures, and constraints," *Phys. Rept.* **537** (2014) 91–121.

[256] A. D. Sakharov, "Violation of CP Invariance, c asymmetry, and baryon asymmetry of the universe," *Pisma Zh. Eksp. Teor. Fiz.* **5** (1967) 32–35. [Usp. Fiz. Nauk **161**, (1991) 61].

[257] P. Mitropoulos, "Right-handed sneutrinos as asymmetric DM and neutrino masses from neutrinophilic Higgs bosons," *JCAP* **1311** (2013) 008.

[258] J. A. Harvey and M. S. Turner, "Cosmological baryon and lepton number in the presence of electroweak fermion number violation," *Phys. Rev.* **D42** (1990) 3344–3349.

[259] S. Chang and L. Goodenough, "Charge asymmetric cosmic ray signals from dark matter decay," *Phys. Rev.* **D84** (2011) 023524.

[260] M. T. Frandsen and S. Sarkar, "Asymmetric dark matter and the Sun," *Phys. Rev. Lett.* **105** (2010) 011301.

[261] M. R. Buckley and S. Profumo, "Regenerating a symmetry in asymmetric dark matter," *Phys. Rev. Lett.* **108** (2012) 011301.

[262] L. Bergstrom, "Nonbaryonic dark matter: Observational evidence and detection methods," *Rept. Prog. Phys.* **63** (2000) 793.

[263] G. Belanger, A. Pukhov, and G. Servant, "Dirac neutrino dark matter," *JCAP* **0801** (2008) 009.

[264] L. Feng, S. Profumo, and L. Ubaldi, "Closing in on singlet scalar dark matter: LUX, invisible Higgs decays and gamma-ray lines," *JHEP* **03** (2015) 045.

[265] S. Profumo, M. J. Ramsey-Musolf, and G. Shaughnessy, "Singlet Higgs phenomenology and the electroweak phase transition," *JHEP* **08** (2007) 010.

[266] LUX Collaboration, D. S. Akerib *et al.*, "First results from the LUX dark matter experiment at the Sanford Underground Research Facility," *Phys. Rev. Lett.* **112** (2014) 091303.

[267] E. Del Nobile, M. Nardecchia, and P. Panci, "Millicharge or decay: A critical take on minimal dark matter," *JCAP* **1604** (2016) 48.

[268] L. Di Luzio *et al.*, "Accidental matter at the LHC," *JHEP* **07** (2015) 074.

[269] M. Cirelli, N. Fornengo, and A. Strumia, "Minimal dark matter," *Nucl. Phys.* **B753** (2006) 178–194.

[270] B. Ostdiek, "Constraining the minimal dark matter fiveplet with LHC searches," *Phys. Rev.* **D92** (2015) 055008.

[271] M. Cirelli, F. Sala, and M. Taoso, "Wino-like minimal dark matter and future colliders," *JHEP* **10** (2014) 033.

[272] R. Essig, "Direct detection of non-chiral dark matter," *Phys. Rev.* **D78** (2008) 015004.

[273] M. Beneke *et al.*, "Relic density of wino-like dark matter in the MSSM,".

[274] N. G. Deshpande and E. Ma, "Pattern of symmetry breaking with two higgs doublets," *Phys. Rev.* **D18** (1978) 2574.

[275] R. Barbieri, L. J. Hall, and V. S. Rychkov, "Improved naturalness with a heavy Higgs: An alternative road to LHC physics," *Phys. Rev.* **D74** (2006) 015007.

[276] R. Essig *et al.*, "Working group report: New light weakly coupled particles," in *Community Summer Study 2013: Snowmass on the Mississippi* (*CSS2013*) *Minneapolis*, MN, USA, July 29–August 6, 2013.

[277] J. Redondo and M. Postma, "Massive hidden photons as lukewarm dark matter," *JCAP* **0902** (2009) 005.

[278] J. Jaeckel, J. Redondo, and A. Ringwald, "Signatures of a hidden cosmic microwave background," *Phys. Rev. Lett.* **101** (2008) 131801.

[279] H. An, M. Pospelov, and J. Pradler, "New stellar constraints on dark photons," *Phys. Lett.* **B725** (2013) 190–195.

[280] H. An *et al.*, "Direct detection constraints on dark photon dark matter," *Phys. Lett.* **B747** (2015) 331–338.

[281] T. Hambye, "Hidden vector dark matter," *JHEP* **01** (2009) 028.

[282] M. Vogelsberger, J. Zavala, and A. Loeb, "Subhaloes in self-interacting galactic dark matter haloes," *Mon. Not. Roy. Astron. Soc.* **423** (2012) 3740.

[283] M. Maggiore, *A Modern Introduction to Quantum Field Theory*. Oxford University Press, (Oxford Series in Physics, 12. ISBN 0 19 8520735), 2005.

[284] M. D. Schwartz, *Quantum Field Theory and the Standard Model*. Cambridge University Press, 2014.

[285] G. Krnjaic, "Probing light thermal dark-matter with a higgs portal mediator," (2015).

[286] H. Baer and X. Tata, *Weak Scale Supersymmetry: From Superfields to Scattering Events*. Cambridge University Press, 2006.

[287] M. Dine, *Supersymmetry and String Theory: Beyond the Standard Model*. 2007.

[288] J. Edsjo, *Aspects of neutrino detection of neutralino dark matter*. PhD thesis, Uppsala U., 1997.

[289] D. Pierce and A. Papadopoulos, "The complete radiative corrections to the gaugino and Higgsino masses in the minimal supersymmetric model," *Nucl. Phys.* **B430** (1994) 278–294.

[290] L. Bergstrom *et al.*, "Gamma rays from Kaluza-Klein dark matter," *Phys. Rev. Lett.* **94** (2005) 131301.

[291] S. M. Carroll, *Spacetime and Geometry. An Introduction to General Relativity.* 2004.

[292] Hitomi Collaboration, F. A. Aharonian, H. Akamatsu, F. Akimoto, S. W. Allen, L. Angelini, K. A. Arnaud, M. Arnaud, H. Awaki, M. Axelsson *et al.*, "Hitomi constraints on the 3.5 keV line in the Perseus galaxy cluster" (2016), arXiv: astro-ph.HE 1607.07420, http://adsabs.harvard.edu/abs/2016arXiv160707420H, Provided by the SAO/NASA Astrophysics Data System.

Index

I

J

K

L

M

N

P

Q

R

11244313R00156

Printed in Great Britain
by Amazon